EMBODIED RESEARCH METHODS

Sara Miller McCune founded SAGE Publishing in 1965 to support the dissemination of usable knowledge and educate a global community. SAGE publishes more than 1000 journals and over 800 new books each year, spanning a wide range of subject areas. Our growing selection of library products includes archives, data, case studies and video. SAGE remains majority owned by our founder and after her lifetime will become owned by a charitable trust that secures the company's continued independence.

Los Angeles | London | New Delhi | Singapore | Washington DC | Melbourne

EMBODIED RESEARCH METHODS

TORKILD THANEM & DAVID KNIGHTS

SAGE

Los Angeles | London | New Delhi
Singapore | Washington DC | Melbourne

SAGE

Los Angeles | London | New Delhi
Singapore | Washington DC | Melbourne

SAGE Publications Ltd
1 Oliver's Yard
55 City Road
London EC1Y 1SP

SAGE Publications Inc.
2455 Teller Road
Thousand Oaks, California 91320

SAGE Publications India Pvt Ltd
B 1/I 1 Mohan Cooperative Industrial Area
Mathura Road
New Delhi 110 044

SAGE Publications Asia-Pacific Pte Ltd
3 Church Street
#10–04 Samsung Hub
Singapore 049483

Editor: Kirsty Smy
Editorial assistant: Martha Cunneen
Production editor: Tanya Szwarnowska
Copyeditor: Jane Fricker
Proofreader: Neil Dowden
Marketing manager: Susheel Gokarakonda
Cover design: Francis Kenney
Typeset by: C&M Digitals (P) Ltd, Chennai, India
Printed and bound by CPI Group (UK) Ltd,
Croydon, CR0 4YY

MIX
Paper from responsible sources
FSC® C013604
FSC www.fsc.org

First published 2019

Library of Congress Control Number: 2018958338

British Library Cataloguing in Publication data

A catalogue record for this book is available from the British Library

ISBN 978-1-4739-0440-8
ISBN 978-1-4739-0441-5 (pbk)

To the memory of Carl Lundevi

CONTENTS

ABOUT THE AUTHORS

Torkild Thanem is Professor of Management and Organization Studies at Stockholm Business School, Stockholm University, Sweden. He is an Associate Editor of the journal *Organization*, and a former Associate Editor of *Gender, Work & Organization* (2009–2018). Torkild's current research focuses on two main areas: fitness and wellness in corporate performance cultures; and the lived practices, experiences and politics of trans people. His most recent work has been published in *Business Ethics Quarterly, Harvard Business Review* and *Organization: The Critical Journal of Organization, Theory & Society*.

David Knights is Professor in the Department of Organization, Work and Technology at the Lancaster University Management School, UK. David's recent empirical research has been on academics and veterinary surgeons and his theoretical interests are broadly concerned with the body and ethics, gender and diversity, identity and power, and management and leadership. His latest work has been published in the *International Journal of Management Reviews, Human Relations* and *Gender, Work & Organization*, and he recently co-edited the book *"Leadership Matters?": Finding Voice, Connection and Meaning in the 21st Century* (with Chris Mabey, published by Routledge in 2017). David co-founded the journal *Gender, Work & Organization* and was its co-editor from 1994 to 2016.

PREFACE

WHY WE WROTE THIS BOOK AND WHY YOU MIGHT BE READING IT

After years of treating the human body in a disembodied manner or ignoring it altogether, the body has become a thriving topic in a number of fields and disciplines across the social sciences and humanities. From philosophy and intellectual history, via anthropology and sociology to pedagogy, cultural studies, organization studies and several other areas, the body has emerged as an increasingly important subject and object of knowledge. Though this should not inflate our expectations of more embodied forms of engagement, the body is now explored from a wide variety of angles in order to understand its involvement in social action and interaction, and examine its significance for a range of different phenomena: culture and class, power and identity, gender and sexuality, age and ethnicity, health and illness, impairment and disability, work and consumption, learning and knowledge are but some of the subject matters benefiting from this growing concern with the body.

The body has been a primary concern in our own research too. Whether we have studied corporate performance cultures, transgender communities or the veterinary profession, it has been impossible to ignore the embodied nature of life and work in these settings or avoid reflecting about how we are involved with our own bodies in the research we do. Torkild would be the first to admit that his research on transgender is shaped by his own transvestism, and David would be hard pressed to deny that his concerns with masculinity were made more vivid by his own embodied experience of infertility as a young man, though, admittedly, we have been able to do this work from a privileged position as white middle-class men.

At the same time, the existing textbook literature on qualitative research methods has given little attention to the body and embodiment. When we randomly examined 20 contemporary texts on methods and methodology in the social sciences we found that only a handful of them made any reference to the body or embodiment. And when we made a systematic search for the terms 'body', 'bodies' and 'embodiment' across the entire population of social science textbooks on methods and methodology, we were only able to identify a handful more texts that discussed the body, few of which dealt with it in any elaborate or systematic manner. This is a problem, not least because textbooks provide the first encounter that most students have with the subject and practice of research method.

In this book, we seek to advance research methods that exploit the body's capacity to generate knowledge, craft lively accounts and create fleshy concepts which enrich our understanding of how people live, work and interact with their bodies in the social world. To do so, we engage with scholarly approaches that appreciate how the body is both a medium and transformed outcome of knowledge, and we discuss methods, practices and personal experiences which profoundly involve our bodies in the research process; in generating and analysing empirical material, in reflecting about the work we do as researchers, and when turning our research into written text. Research is no disembodied matter of the mind alone. The mind is a bodily organ, our bodies are embedded in social settings, and we all use our bodies in our striving to understand social life.

Since all research and all aspects of social life are in some sense embodied, we trust that students and academics across a number of different fields and disciplines will find this book valuable. Although students picking up this book may be unlikely to do so as part of a separate course on embodied research methods, it should be relevant for almost anyone seeking to broaden their repertoire of qualitative research methods, and anyone keen to gain a fresh approach to key problems and issues in qualitative research.

Most of the techniques, procedures and activities that we engage with here may be familiar from standard methods books and conventional discussions of qualitative method. But unlike such treatments, we have sought to explore the embodied aspects of these practices, whether we are dealing with interviewing or participant observation, photography or diaries, access negotiation, analysis, or the actual process of writing. And more importantly, we have sought to unpack their potential for investigating the bodily features of social life.

Doing so, we have avoided the temptation to offer neat recipes, frameworks and toolboxes. Rather than glossing over the many difficulties and frustrations that pervade all research processes, we have attempted to bring them out in the open by sharing our own experiences and those of other academics. There is no best way of conducting embodied research, or any research for that matter. Whichever way you go about it, you will run into problems. Without promising that this book will help anyone perfectly resolve the many embodied snags and troubles of the research process, we hope it will help you engage with them and recognize that you are not alone in all of this.

WHAT YOU WILL FIND IN THIS BOOK

To further demonstrate the relevance of embodied research methods and indicate their potential, we open this book with an introductory chapter on the embodied turn in the social sciences. The chapter offers an overview of key themes in embodied research and research on the body, emphasizing the body's involvement in social interaction, politics and work, how the body is both a subject and object of knowledge, and how all aspects of our everyday world are lived and experienced in and through the body. At the end of the chapter we introduce some of the ethical, political and meta-theoretical issues that emerge when doing embodied research.

Despite this embodied turn, embodied research and embodied research methods remain a contested issue. Even some of the contemporary social science research on the body may be critiqued for approaching the body in a distanced and disembodied way. In order to understand why the body has a long history of neglect and why embodied methods continue to be under attack within the social sciences, we will proceed in Chapter 1 by examining the disembodied heritage that has underpinned western intellectual thought and the main streams of social science research. In this part of the chapter we will focus on how the body has been viewed by the most central figures in the history of philosophy and sociology, namely Plato, Aristotle, Descartes, Kant and Durkheim. Readers looking for further nuance and elaboration may want to look up the notes we have added in these passages; other readers may want to skip the notes. We end this chapter by considering how the contributions and thinkers we have discussed offer disembodied answers to questions about the body.

In Chapter 2, we discuss the main features of embodied methodology along with some of the key problems they confront us with. To do so, we will draw on concrete examples from feminism, phenomenology and other research traditions that are underpinned by embodied methodology. We will also engage with meta-theoretical questions of what the body is, what we can know about the body and what kinds of methods one might use in order to investigate it. Doing so, we aim to tease out what we mean by the lived reality of embodiment and discuss how well the body can be known and represented. We will also start considering how we may use our bodies to access the body as an object of knowledge.

In the remaining chapters of the book we shall look more closely at how various methods and approaches may be used in embodied research. Since research is a temporal though not entirely linear process, we shall do our best to consider all the main phases that we normally encounter during a research process: from preparation to dissemination.

In Chapter 3, we consider the initial and preparatory stages of doing embodied research. Here, we will especially deal with the process of planning and initiating an embodied research project, and the challenge of negotiating access to empirical settings that lend themselves to such projects. But we start the chapter by considering the bodily sensation – the *itch* – which often prompts people to investigate a new research problem in the first place.

In Chapters 4 to 7, we focus on techniques, procedures and methods that are usually associated with the process of data generation. In Chapter 4, we discuss how ethnographic fieldwork methods are not limited to observing other people and their bodies but allow us to immerse ourselves so much in the settings we study that we may start to feel, experience and understand what it may be like for ordinary participants to live and work there.

In Chapter 5, we move on to see how interviews and conversations may be utilized to generate bodily and embodied material about the people we talk with. While interviews and conversations are typically appreciated because they allow researchers to generate textual data and narrative material, we stress that they are face-to-face encounters which offer good opportunities for observing participants' facial expressions, gestures and bodily demeanour, for sensing their moods and feelings as well as their tastes and their approach to life and work.

Chapter 6 considers how embodied material may be recorded and documented by examining visual and biographical methods. Whereas visual methods such as photography and video enable researchers to record and share what bodies look like, biographical methods such as journals and diaries enable participants and researchers to probe into their own intimate spheres and reflect about their own embodied feelings and experiences. We shall not forget, however, that the pen and the notebook are extremely convenient devices that allow researchers to jot down observations, conversations and impressions in an instant, without encroaching too much on our bodily involvement.

Chapter 7 discusses how we may begin to interpret, analyse and make sense of embodied empirical material without reducing bodily feelings, lively gestures and vivid accounts to neat concepts and categories. Instead of rejecting analysis as a disembodied enterprise of distanced decomposition and reduction, we propose to 'bugger' analysis in a way that embraces its anatomical roots, scars and pores, and thrives on a monstrous intercourse of theoretical concepts, incomplete findings, visceral experiences and personal troubles.

In Chapter 8, we address the ongoing and embodied process of writing, and the possibilities of disseminating embodied research. We first engage with feminine and masculine traditions of writing to see how embodied research can be pursued through vivid accounts that allow us to flesh out and vibrate the social issues we are concerned with against the surfaces of our bodies and the depths of our lived experiences. We also consider the physical nature of writing. We close this chapter by discussing the challenges and joys of publishing and presenting embodied arguments and research findings in academic and popular outlets, and the chances of doing so in ways that retain at least some of the viscerality experienced while doing embodied research.

Like all research, working with embodied methods actualizes a number of ethical questions and dilemmas. We shall engage with them throughout the book, as they emerge in connection with specific procedures, methods and approaches.

However, embodied research has also been the target of criticism: as if, by nature, it is an apolitical project which fails to understand big political problems yet facilitates the proliferation of an unjust and irresponsible politics based on personal tastes and experiences, alternative facts and the egotistic interests of individuals and identity groups.

While this book draws on Marxist, feminist and other traditions of scholarship which are both embodied and profoundly political, we shall return to discuss the political implications and opportunities of embodied research at some length in Chapter 9.

In our final and concluding chapter, we summarize the key lessons and challenges of working with embodied research methods, and encourage future scholarship to cultivate an embodied imagination through engaged and embodied research projects that flesh out the lived reality of social problems.

Even though method and methodology may be interesting topics in their own right, our primary intention with this book is to help you tackle concrete problems and challenges when doing embodied empirical research. At the end of each chapter, we therefore provide a set of questions and exercises that give you the opportunity to practise the various methods and approaches that you have been reading about.

ACKNOWLEDGEMENTS

When we started this book project back in January 2014 we were confident that we only needed one and a half years to complete it. We had already published a few joint papers on embodied issues and methods, and we were hoping that most of that material could go into the book without requiring much tweaking. How wrong we were. Working on this book, we soon realized that we had to revise many of the assumptions and arguments from our previous research. Some of them even had to be rejected. Thus, none of our previous publications proved appropriate for a 'cut-and-paste' treatment, and we are grateful to the Sage editor Kirsty Smy and her editorial team for the patience and trust they have shown us during this process.

Over the past few years we have presented some very early drafts of short bits and pieces at various conferences and seminars. Torkild presented some rudimentary thoughts at a seminar at Essex University in November 2015, and we did a joint presentation on Merleau-Ponty's embodied phenomenology and politics at the Asia and Pacific Rim Organization Studies conference at the University of Technology in Sydney in December 2015. At the Gender, Work & Organization conference at Keele University in June 2016, Torkild presented some of the ideas which later made up key sections of Chapter 2 on embodied methodologies. We would like to thank the organizers of these events and everybody who engaged us in discussion during and after these presentations.

Like the body, money is no immaterial matter, and we would therefore like to acknowledge two funding bodies that sponsored this book project. An international visiting grant from Handelsbankens Forskningsstiftelser enabled us to meet face-to-face while working on this book and to buy other books that helped us develop our argument. A project grant from the Swedish Research Council enabled Torkild to devote more time to writing, and to conduct a study on corporate performance culture in a Swedish sportswear company, which in no meagre degree has shaped his experience of working with embodied research methods. Special thanks are due to the many research participants who became part of his study, especially those in the sportswear company, whose names, for confidentiality reasons, cannot be made public here; without you, I would not have been able to immerse myself in the craft of embodied methods, and consequently this book may never have been completed; the CEO and the HR Manager deserve a special acknowledgement for granting me such generous access, as does my colleague Carl Cederström, who helped me negotiate access to the company.

Furthermore, we would like to recognize a number of colleagues, friends and students who, sometimes inadvertently, have helped us develop our thoughts on the topic. David acknowledges his students from teaching a third-year course on work, industry and society in his formative years, but also his research students, most of whom became academics. Deborah Kerfoot, with whom he first began writing about the body and thinking in terms of embodied research, deserves a special mention, but all his other colleagues had an impact in those early years, especially Hugh Willmott who worked with him in developing the course.

Torkild would like to thank the many master's students who have attended his qualitative methods class at Stockholm Business School between 2014 and 2018, the group of doctoral students who took the course on qualitative research methods in the autumn of 2016 and his colleague Maria Frostling, with whom he co-taught the doctoral course. He also thanks Raoul Galli for sharing his anthropological expertise and for being such a generous and humble conversation partner on all issues ethnographic. Finally, Torkild thanks Louise, who continues to be his best critic, interrogator and support.

INTRODUCTION

The Embodied Turn in Social Science Research

BODIES EVERYWHERE

Despite the striking absence of books dealing with embodied research methods, the social sciences have witnessed a huge interest in the body over the past few decades. Contributions in a variety of fields and disciplines have explored the body's involvement in work and politics, the embodied aspects of action, social interaction and meaning-making, the lived and reflexive nature of embodiment, and how the body is a target of knowledge yet actively involved in its creation.

To appreciate the importance of refining our methods for investigating bodies and embodiment in social life, we shall use the bulk of this chapter to map out the main themes of this embodied turn. At the end of the chapter we shall raise some of the ethical, political and meta-theoretical questions that emerge when doing embodied research.

BODY POLITICS

Through the embodied turn in social science research, it has become widely recognized that the body is a site and medium of desire and transgression as well as an object of discursive construction and a target of social differentiation, discrimination, discipline and normalization. As Simone de Beauvoir (2011 [1949]) argued in *The Second Sex*, girls and boys initially 'have the same interests and pleasures' and comprehend the world 'not through their sexual parts' but with their eyes and hands (p. 330); it is through our continuous engagement with others that we, and our relationships with our bodies, become sexually and socially disciplined and differentiated.

In a different vein, Michel Foucault (1979 [1977]) insisted that the body became a target of disciplinary control only after it was conceived of as a subject of sexual desire and transgression by powerful regimes of economic and scientific discourse. In the *History of Sexuality* he argued that 'the Malthusian couple', 'the hysterical woman' and 'the masturbating child' (p. 105), all of which preoccupied western scientists and public officials in the 19th century, could only be subject to industrial, medical and pedagogical discipline once their sexual desires had become knowable through demographic and sexological investigation.

Subsequently, several studies have shown how these relationships are played out in a variety of social settings; through practices of social classification and discrimination that differentiate and rank bodies, and through technological, consumerist and other forms of intervention and regimentation that are employed to manage and improve our bodies. Indeed, some of the feminist scholarship that pioneered the social and cultural understanding of the body emphasized how women employ cosmetic surgery, dieting and physical exercise in order to conform to prevailing norms of beauty and bodily appearance (e.g. Morgan, 1991), whilst stressing that these practices are 'the logical' response to 'anxieties and fantasies fostered by our culture' (Bordo, 1993, p. 15).

Related contributions have been made in our own field of organization studies. According to feminist research in this area, male-dominated work organizations take the male body for granted but deem the female body leaky and problematic (Acker, 1990). Such tendencies reduce the female body to an object of male heterosexual desire which is vulnerable to sexual harassment (Rogers and Henson, 1997), devalued through gendered pay gaps (Misra and Murray-Close, 2014) and aestheticized through stereotypically feminine dress codes and beauty norms (e.g. Tyler and Abbott, 1998). As women in such organizations are simply expected to discipline their bodies according to prevailing gender norms, it has been argued that women who resist or fail to do so risk stigmatization, marginalization and impaired career prospects (e.g. Trethewey, 1999; Haynes, 2012).[1]

However, the political potential and agency of the body has by no means been ignored. Judith Butler's (1990) theory of gender performativity did not merely explain how bodies conforming to the dominant discourse of binary gender norms become intelligible, legitimate and liveable while those who transgress it become unthinkable, abject and unliveable. Butler (1990, 2004) has been especially memorable in stressing how people make gender trouble through embodied acts and expressions that challenge the binary opposition between masculinity and femininity as well as the assumed correspondence between social gender and biological sex (see, e.g., Halberstam, 1998; Na and Choo, 2011; Thanem and Wallenberg, 2016).[2]

The body's transgressive potential has been further explored through the work of philosophers such as Gilles Deleuze and Luce Irigaray. Deleuze and Guattari's (1984 [1972], 1988 [1980]) concepts of 'the body without organs' and 'becoming-woman' have been utilized to consider how the libidinal forces of transgender bodies might precede, exceed and subvert the binary sexual-social order (see, e.g., Thanem, 2004, 2006, 2011; Linstead and Pullen, 2006). In contrast, Irigaray's (1985 [1977]) notion of the female body's essential difference has been mobilized to explore the possibilities of radical femininity within and around organizations (see, e.g., Vachhani, 2012; Phillips et al., 2014; Pullen, 2018).

More empirically oriented studies have drawn attention to the body's concrete involvement in political action and resistance, and the tensions and contradictions of using the body as an instrument of political protest. For example, research on the suffragette movement has highlighted the physical and psychological toughness of suffragette activists: the courage and steadfastness involved in displaying suffragette colours in public at the risk of physical attack from political opponents and the police; the strength and agility exercised when climbing rooftops to protest against anti-suffragette meetings in adjacent buildings; and the tenacity demonstrated by suffragettes who went on hunger strike during imprisonment (Parkins, 2000).

Equally embodied findings have been reported from contemporary settings. This includes studies of naked protest by male activists at the World Social Forum (Sutton, 2007), research on topless culture jamming by Femen activists protesting the sexploitation of women's bodies (Reestorff, 2014), and accounts of self-mutilating protest, suicide attempts and hunger strikes by migrants incarcerated in Europe (Puggioni, 2014).[3]

EMBODIED ACTION, INTERACTION AND MEANING-MAKING

These examples are by no means exceptional. In a general sense, all social action and interaction is embodied. After all, it is through our bodies that our actions, subjectivities and relations with others are realized, materialized and made visible (e.g. Shilling, 1993; Crossley, 1995). Erving Goffman (1959, 1961) addressed this point when highlighting the social function people perform when managing impressions through dramaturgical techniques of face-work. Pierre Bourdieu (1977, 1984) did so when stressing how social structures are embodied, reproduced and habituated through the distinctive tastes, mannerisms and accents that characterize different social groups.

Although western norms of bodily conduct and appearance may be more relaxed now than when Goffman and Bourdieu wrote about these issues, our body language and bodily appearance remain subjects and objects of communication, meaning-making and identity expression. Even at a time when everyday activities are enabled and mediated through social media, email and other forms of technology, much of our social interaction continues to be constituted through face-to-face encounters between our bodies and those of other people. We still seek to tell others who we are through our bodies, and others make sense of who we are by observing our bodies and attributing meaning to them.

These claims find considerable support in empirical research. For instance, a recent study has suggested that women elite leaders may seek to express a respectable business identity by adopting a moderately feminine style of clothing and personal grooming that is neither deemed too 'girly' nor too masculine (Mavin and Grandy, 2016). More blatantly, it has been argued that 'beauty pays': research by the economist Daniel Hamermesh (2013) suggests that, in the US, 'attractive' men are likely to earn 4% more than average-looking men with similar education and experience, and that 'ugly' men are likely to earn 13% less than men with average looks. Other studies suggest that firms run by male CEOs with wide and stereotypically masculine faces perform better financially (Wong et al., 2011), that voters prefer physically attractive candidates (Efran and Patterson, 1974; Berggren et al., 2010), and that black senior executives with so-called baby-face features are more successful because they ostensibly seem less threatening to white people (Livingston and Pearce, 2009).

While we personally would tend to privilege social science contributions which challenge established modes of thought and practice (Giddens, 1984), perhaps we should not be too surprised that much of the empirical work in this area tends to confirm commonsense notions about the role of physical appearance at work.

THE WORKING BODY

Our bodies are indeed central to the work we do. Karl Marx recognized this in a more profound way, as he maintained that the body is both a subject and object of labour (see also Wolkowitz and Warhurst, 2010). In the *Grundrisse* he argues that labour 'exists ... as a resource in the bodiliness of the worker' (1973 [1857–8], p. 298). Returning to this point in the first volume of *Capital*, he elaborates that labour involves moving 'the natural forces that belong to [our] own body, [our] arms and legs, head and hands' (1976 [1867], p. 283), and he emphasizes that we enjoy work that involves 'the free play of [our] own physical and mental powers' (p. 283).

When he was younger, Marx had warned readers of the consequences should we be turned into passive objects cut off from the free play of our minds and bodies. In the *Economic and Philosophic Manuscripts of 1844*, he argues that the estranged labour characterizing industrial capitalism 'mortifies [the] body and ruins [the] mind' because labour then becomes a coercive and alienating activity that 'belongs to another' (2009 [1959], p. 30). As labour becomes an externalized object of which we are no longer in control, our bodies too are transformed into an object that we are alienated from.[4] For the objectified working classes that Marx wrote about, the pain of labour was excruciating; their lives and bodies were marked by 'massive hunger, sores, disease, airlessness, and exhaustion' (Scarry, 1985, pp. 170–1).

Without ignoring that many workers in farming, construction and other industries continue to live and work under similarly appalling conditions, the current pains and sufferings of working life are multifarious. While this is documented through studies of stress, accidents and other work-related health problems in the vast fields of occupational medicine, work psychology and health sociology (see, e.g., Theorell and Karasek, 1996; Hyde et al., 2006; Cooper and Dewe, 2008; Lu, 2009), what characterizes the contemporary objectification of the working body is perhaps more apparent in front-line service work and other occupations that involve aesthetic and emotional labour. As Arlie Hochschild (1983) showed in her classic study of Delta Air Lines cabin crew, women flight attendants were systematically trained to suppress negative feelings, keep eye contact with customers and lay on a big smile regardless of their inner feelings. Later studies have extended these insights, observing how women flight attendants are subject to stricter norms of body weight, body shape, hairstyle and grooming than their men colleagues (see, e.g., Tyler and Abbott, 1998; Hancock and Tyler, 2000).

In these lines of work, people are expected to transform their bodies into an extension of the corporate brand by manufacturing an emotional, aesthetic and bodily display that communicates company values (see also Warhurst and Nickson, 2009). Although there is an expanding industry of hairdressers, manicurists and other 'body workers' who help people sculpt a bodily appearance deemed appropriate for interactive service work (Wolkowitz, 2006), individuals are largely made responsible for doing the emotional and aesthetic 'body work' necessary to objectify their bodies in this way. Consequently, the body becomes the subject of its own objectification.

THE REFLEXIVE BODY

It may be argued that this is the essence of embodied reflexivity: that the body is both subject and object of the same actions (Crossley, 2006b). As body-subjects, we turn our bodies into objects of our own bodily practices: perceiving, feeling and thinking about our own bodies, looking at them, and comparing them to other people's bodies. We feed ourselves and starve ourselves, and we groom, dress and exercise ourselves, not merely to fulfil physical needs or because these activities have become habits or routine, but also to say something about who we are and who we are in relation to others. We maintain, modify and change our bodies in order to fit in or to be unique (Crossley, 2005), in order to do our moral duty to society (Crossley, 2006a; Zanker and Gard, 2008), or in order to suggest that life, and society, can be different (Bigo and Islam, 2018).

These issues of embodied reflexivity have been well documented by empirical research on body techniques and body modification: straight and gay men's pursuit of a muscular body has been linked to a striving for social status and self-esteem (Pope et al., 2000; Duncan, 2010); tattooing and piercing have been associated with efforts of self-creation rather than rebellion and marginalization (Sweetman, 1999); self-harming behaviours such as anorexia and binge-drinking among young women have been found to reproduce as well as resist idealized forms of femininity (Day, 2010); furthermore, the embodiment of a fat identity has been seen as a 'subversive ... form of political resistance' (Gailey, 2012, p. 122).

THE LIVING BODY

For Maurice Merleau-Ponty (1968 [1964]) the body's simultaneous subjectivity and objectivity was more a question of *reversibility* than reflexivity. Through the ideal example of folding and pressing his own hands, he emphasized the 'touching of the touch, when my right hand touches my left hand'. As 'the "touching subject" passes over to the rank of the touched' (pp. 133–4), Merleau-Ponty insisted that this corporeal reversibility cannot be grasped through 'double sensation' because the hands' subjectness and objectness cannot be sensed simultaneously: 'When I press my hands together, it is not a matter of two sensations felt together as one perceives two objects placed side by side, but of an ambiguous relationship in which both hands alternate the roles of "touching" and being "touched"' (1962 [1945], p. 106).

While this illustrates the mind's inability to capture the lived experience of the body from a distance, Merleau-Ponty maintained that the mind is neither separate from the body nor capable of ordering the body and its perceptions (1962, p. xii). Rather, the body both thinks and perceives (p. xii), and 'the perceiving mind is an incarnated mind' (p. 3). The body lives, and the mind is a living organ within the body, yet the living body exceeds our capacity to fully sense it and understand it.

THE KNOWING BODY

Notwithstanding these limits, what we know and how we learn is shaped and enabled by our bodies. Far beyond Merleau-Ponty's philosophical investigations, this point has found support in empirical studies of adult professionals and school children. Research on expert anaesthetists argues that nods and hand movements made between colleagues in an operating theatre are not merely perceived as bodily gestures but interpreted as offers and requests for assistance (Hindmarsh and Pilnick, 2007). In a similar vein, research on primary school mathematics teaching suggests that first-grade school children's knowledge of geometry is supported by and expressed through bodily gestures, which make abstract geometrical concepts such as circles and squares concrete (Kim et al., 2011).

The body, then, is an active subject of learning and knowledge creation as well as an expressive medium of gestures and facial expressions; it is a cognitive medium of thinking as well as a sensory medium of emotion and lived experience. While we cannot walk or talk without our bodies, we also cannot read, write or think without them. According to the gestural theory of language, these capacities are deeply ingrained in human evolution; it is unlikely that human language would have developed in the first place had not humans become bipedal creatures whose hands could be used for other purposes than walking (Corballis, 2002).

Either way, we cannot exist and act without our bodies, and we cannot imagine how we might do research without them. Yet, there is far more research about bodies than there are methodological discussions about how we use and rely on our bodies in our research. With few exceptions, most social scientists and methods books give the impression that research is a disembodied and purely cognitive process. But surely, every aspect of research engages the body in some way, whether we are observing others by looking, interviewing others by listening and talking, or seeking to share their worlds by touching the artefacts that occupy their surroundings, tasting the food they eat and absorbing the smells filling their spaces.

Over the past few years, a number of social scientists have begun exploring how we might use our bodies as instruments in our research and reflect about the various ways in which our bodies are involved in the research process (e.g. Engelsrud, 2005; Thanem and Knights, 2012). Efforts have been made to develop 'sensory scholarship' and 'sensory ethnography', which encourages us to use all our bodily senses in our research (Stoller, 1997; Pink, 2015). Strong arguments have also been put forward for a 'sociology of flesh and blood', which urges researchers to become bodily immersed in the empirical settings that draw our attention (Wacquant, 2015).

Others have developed explicitly embodied tools, techniques and methods: videotaped smell interviews, paper strips and audio smell diaries have been employed to grasp the somatic work that people engage in to deal with sensory experiences in work organizations (Riach and Warren, 2015). Diaries have proved useful in recording the visceral and emotional experiences of bodily activities such as long-distance running (Allen-Collinson, 2011, 2013); photographs taken by research participants have been used to elicit conversation about their embodied experience of work (Slutskaya et al., 2012).

Some such methods have been particularly effective in illuminating the differential experiences that people have from life and work. In disability studies there is a long tradition of investigating physical barriers in the material environment by embodying the research field as a wheelchair user, as visually impaired and as hearing impaired (see, e.g., Imrie, 2004). In certain strands of feminism there is a rich tradition of embodied writing, which offers vivid accounts of the pain and suffering caused by ongoing experiences of sexism and racism, and passionate calls for radical political action (see, e.g., Cixous, 1976 [1975]; Moraga and Anzaldúa, 1983; Pullen, 2018). Regardless of their differences, all these contributions alert us to the fact that we use our bodies when conducting research.

META-THEORETICAL QUESTIONS WE MUST ASK OURSELVES WHEN DOING EMBODIED RESEARCH

That we necessarily use our own bodies when researching other people and their bodies raises a number of meta-theoretical questions: *ontological* questions about what the body is; *epistemological* questions about what we can know about it; and *methodological* questions about what kinds of methods, tools and techniques we might utilize in order to investigate it.

Are we able to know the truth about the body, or can we 'merely' hope to generate meaningful accounts and convincing interpretations of bodily expressions, practices and experiences? Is it possible to study bodies in a disembodied fashion, or does embodied research require particular methods and styles of writing by which we express the visceral experience of being fully immersed in the social settings we study? To what extent are the bodies we study constructed by the concepts and ideas we use to make sense of them, to what extent do they exist independently of such constructs, and how far do embodied feelings, experiences and expressions exceed our efforts to know them? Are our bodies determined by natural, historical and social laws, are we able to take care of and even re-create our bodies as we wish, or are they indeterminate products of a multiplicity of material and discursive forces?

We will not be able to offer comprehensive answers to these questions in this book. However, we will return to them in different chapters when discussing specific methods and research traditions. Here, it may suffice to make the following comment: that even though we are unconvinced that the living body can be fully known, represented or managed, the massive amount of social science research on the body encourages us to do what we can to advance the prospects for learning how people experience, make sense of, act and interact with their bodies in social life.

ETHICS AND POLITICS IN EMBODIED RESEARCH

Furthermore, embodied research methods actualize a range of ethical and political questions. By investigating other people's bodies, we often touch upon sensitive issues and intimate matters. Does this mean that we encroach upon their personal lives in ways that make them particularly vulnerable? Is it possible to justify invasive research practices on the basis that the anticipated results might help push for change which makes life more liveable for more people? Indeed, how do we engage responsibly with the people who participate in our research? These are tricky dilemmas that we shall return to throughout the book. Instead of trying to squeeze them into simple ethical frameworks of fixed universal duties, common-sense virtues or cost–benefit calculations, we recognize that they require careful consideration and engagement in each case.

The political challenges of doing embodied research raise somewhat different questions, particularly as we are confronted with a political climate that is increasingly characterized by impulsive politicians, nationalist sentiments and aggressive polemics against people who think differently, behave differently and look different. Is embodied research at best a form of self-interested identity politics, and at worst apolitical and indifferent to any other concerns but the personal troubles of the embodied researcher? Do embodied research methods involve a narcissistic preoccupation with the self, which is exclusively committed to the researcher's own feelings and body issues? Are embodied research methods incapable of engaging with larger political issues that concern communities and entire societies?

These are certainly some of the criticisms that have been launched at various forms of embodied research, even before the turn to 'post-truth' politics that is currently undermining democratic institutions across the world. While we shall discuss the politics of embodied research at greater length in Chapter 9, let us stress that simple oppositions between reason and embodiment fail to appreciate the embodied aspects of political action and rational thinking that we have highlighted earlier in this introductory chapter.

Thus, we remain sceptical that inequality, injustice and other social problems can be understood without appreciating how they are played out in the everyday lives of real people with real bodies, across bodily differences in appearance, tastes and sentiments, abilities and disabilities, gender and sexuality, age, race and class. We are also not convinced that genuine political change is possible unless we understand how our small, embodied lives are tied up with big political problems. To appreciate these connections, social scientists need embodied research methods.

Questions and exercises

To get a better sense of what could be learned from this introductory chapter, you may now want to answer one or more of the following questions:

1. Based on your reading of this chapter, what do you consider to be the main contributions generated by previous research on the body in the social sciences?
2. How do these contributions affect how you relate to your own research?
3. How would you define the terms method, methodology and meta-theory? What might it involve to ask an epistemological question about the body? How about an ontological question?

1

DISEMBODIED METHODOLOGY

DISEMBODIED TENDENCIES IN SOCIAL SCIENCE RESEARCH ON THE BODY

It is difficult to date precisely when the body emerged as a distinct area of social science research, but in the 1980s and early 1990s a number of contributions began to appear which sought explicitly to address the body as a topic of social science enquiry. In 1984, the sociologist Bryan Turner published a book titled *The Body and Society*, which was one of the first systematic attempts to develop a sociology of the body; in 1995, he became the first editor of the journal *Body & Society*, which he co-founded with the cultural theorist Mike Featherstone. Although Turner (1984) acknowledges that the body has an independent existence beyond the social and linguistic processes that we employ to make sense of it, he is primarily concerned with the body as a problem of social order and regulation. For Turner, the task of society is to regulate the body, and he devotes a fair amount of his book to demonstrate how this is achieved through the disciplinary and discursive practices of patriarchy, religion and government.[1, 2]

While this has led critics to argue that Turner deals with the body in just as disembodied and distanced a manner as other sociological sub-fields approach their respective subject matters (e.g. Williams and Bendelow, 1998), Turner's approach is not unique in this respect. Despite a number of efforts to embody sociology by engaging with the body as both a subject and an object of social action, knowledge and experience (e.g. Crossley, 1995; Radley, 1995; Williams and Bendelow, 1998), there was a tendency in many other writings to emphasize how the body is constituted as an object of discursive construction and disciplinary control.

An illustrative example of this is the sociologist Deborah Lupton's (1995) book *The Imperative of Health: Public Health and the Regulated Body*. Here, Lupton offers an excellent analysis of how the discourses of 'public health and health promotion act as an apparatus of moral regulation' which privileges rational subjectivities that are capable and willing to manage and control their bodies (p. 158). The book includes an astute reading of contemporary public health policies and health campaigns, which are aimed to encourage people to exercise regularly, keep a healthy diet, refrain from tobacco and limit their alcohol consumption. However, little is made of the fact that public health and health promotion is targeted at real people with real bodies. The book is primarily concerned with analysing the intentions, contents, workings and alleged effects of public health and health promotion discourse rather than examining how ordinary people, who are faced with these discursive constructs, experience, live with and take care of their own bodies in everyday life.[3]

This tendency to investigate the body in a relatively abstract, distanced and disembodied manner coincided with social constructionism becoming a dominant perspective in qualitative research across the social sciences, and the works of Michel Foucault reaching canonical status for many a social scientist devoted to the study of bodies. Even though Foucault (esp. 1990 [1986]) acknowledged that there is a lively and evasive body outside of the orderly realm of discourse, his overall project focused on tracing the discursive and institutional practices through which the body became an object of knowledge and power.

In *The Birth of the Clinic* Foucault (1973 [1963]) showed how the diseased body came to be visualized, inspected and treated within the hospitals that emerged in the late 18th century; in *Discipline and Punish* he demonstrated how the prisoner's body came to be monitored, disciplined and reformed within the panoptic structures of the 19th-century prison (Foucault, 1977 [1975]); and in several of his writings he showed how the bodies of the modern state's citizenry came to be subjectified, policed and managed through biopolitical and governmental practices of schooling and education, public hygiene, social insurance and taxation (see, e.g., Foucault, 1979, 2009 [1977–8], 2010 [1978–9]). Thus, Foucault's project was primarily an attempt to trace the discursive ruptures and transformations that have changed how the body is understood, treated and managed across different historical and institutional contexts rather than investigate the lives and bodies that were worked upon by discourse.[4]

We do not want to exaggerate this tendency to construe the body as a passive object of social and discursive construction, and we shall return to examine the growing mass of embodied methodologies in the next chapter. Also, we do not deny that discourse analysis may be an appropriate tool for examining how the body is constructed, subjectified and ordered through linguistic utterances and institutional practices. Discursive and constructionist studies have made important contributions to challenge dominant notions and subjectifications of the body. Nevertheless, it is important that we point out what is at stake when choosing to work with discursive and constructionist approaches.

One such thing is the meaning and status of reflexivity. Reflexivity has not only been seen as a key feature of embodied research (see, e.g., Crossley, 1995, 2006b), it has also become a catchphrase in discursive and constructionist approaches to qualitative research. While some of the discursive research on the body (e.g. Lupton, 1995) and some of the more influential writings on discourse analysis (e.g. Phillips and Hardy, 2002; Fairclough, 2013) have underlined the reflexive nature of social and discursive practices, there is a peculiar lack of reflexivity in many of these accounts. As we pointed out in the introductory chapter, assuming the body's reflexivity implies that the body is both subject and object of one and the same action (Crossley, 2006b). However, when privileging what a discourse says about the body rather than listening to what the body has to say, it becomes an object rather than a subject of discourse. This makes it possible to understand how the body is constituted by discourse, but it does not shed light on how the various habits, gestures and utterances that we express through our bodies may inform or subvert that discourse.

Again, Lupton's otherwise splendid study may serve as a useful illustration here, even though it was written more than 20 years ago. Whereas Lupton argues that researchers and practitioners need to exercise reflexivity by 'question[ing] the practices, knowledges and belief systems of public health and health promotion, including their own privileging of "health", "knowledge" and "rationality"' (1995, p. 159), there is not any reflection in the book about how she relates to her own body or to the health discourse that she investigates. Moreover, the book offers no accounts to show how other people reflect about their bodies or about their relations to prevailing discourses of health.

DISEMBODIED METHODS TEXTS

This discussion remains significant, perhaps particularly because disembodied versions of reflexivity are deeply ingrained in more recent methodology texts that explicitly pursue a reflexive approach to qualitative research. For instance, Alvesson and Sköldberg's (2009) *Reflexive Methodology*, which has become a bestseller in the genre of social science methods textbooks, offers chapters on a variety of research approaches commonly used by qualitative scholars, including grounded theory, discourse analysis, critical theory and poststructuralism. Interestingly, it is only in a sub-chapter on feminism that they discuss the role of reflection or consider how research may be informed by gendered experiences, feelings and emotions. No attention is given to the embodied and visceral aspects of such experiences, no significant mention is made of the body or embodiment, and the authors make no attempt to reflect about their own experiences from their long research careers.

Aull Davies' (2008) book *Reflexive Ethnography* is another telling example in this context. In arguing for a 'continuing reflexive awareness' (p. 23), Aull Davies acknowledges the fundamental reflexivity of research in general and ethnographic research in particular.

Pointing out that the researcher's presence affects the social reality it is claimed merely to describe, she emphasizes the importance of personal and cultural reflexivity and the need to reflect about sociohistorical context and disciplinary belonging. However, this is largely discussed in cognitive terms, and, like Alvesson and Sköldberg, Aull Davies remains disembodied in her own text.

Such disembodied tendencies are equally apparent when looking at the broader area of textbooks on qualitative research in the social sciences. When we randomly examined 20 contemporary methods and methodology texts, we found that much attention was given to discursive and linguistic matters when employing commonly used techniques such as interviews, observations and focus groups (see, e.g., Jorgensen, 1989; Alvesson and Sköldberg, 2009; Kreuger and Casey, 2015). Only a handful of them made any reference to the body or embodiment. Out of these books, Atkinson et al.'s (2001) 507-page *Handbook of Ethnography* was perhaps the most explicit in this respect, with references to medical sociology and the sociology of the body (p. 182), the body–self relationship and the sensuous body (p. 157), the body being mediated by technology (pp. 264–5), and interpretations of women's bodies in relation to their designation as working class (p. 427). The book also includes references to the embodiment of everyday activities in ethnomethodological research (pp. 118–30) and a claim that embodiment is 'the work of ethnography' (p. 444). On the whole, however, these points are made in passing with regard to specific ethnographic studies, and the handbook offers no discussion of the embodied aspects of actually doing ethnography.

Other books in our sample tended to mention the role of body language in various parts of the research process: Heath (in Silverman, 2004) emphasizes the importance of gestures, bodily comportment and bodily activity in the analysis of face-to-face interaction in video-recorded material, arguing that 'Talk and bodily activity are mutually constituted' to provide meaning (Heath, 2004, p. 275; part of this analysis is reproduced in Silverman [2006, pp. 255–7]). In a broader vein, Bryman (2012) makes 11 references to the body, most of them relating to gender and body language, but no references to embodiment. Of the remaining texts we examined there are only nine references to the body or embodiment. In summary, in well over 8,000 pages of methods and methodology texts, there are only 20 references to either the body or embodiment. It is remarkable that so little reference is made to bodily matters when considering the contingent, contextual and uncertain character of most social science research, which at least partly is an outcome of the obdurate yet vulnerable nature of human bodies.

In order to challenge this cognitivist and disembodied tradition, it is important that we try to understand what has made it possible. What lies behind this marginalization of the body, which Leder (1990) has called the body's 'absent presence' in social science research? In the following pages, we shall therefore seek to trace this tradition through some of the most significant figures in the history of philosophy and into the methodological conventions and research agendas that were pursued as the social sciences emerged as modern academic disciplines. Specifically, we will examine the status of the body in the philosophies of Plato, Aristotle, Descartes and Kant, as well as in the sociology of Emile Durkheim.

THE DISEMBODIED HERITAGE OF SOCIAL SCIENCE RESEARCH

Plato's denigration of the body

As we indicate above, the marginalization of the body goes far back in the history of philosophy, at least to Plato's idealism. In the *Phaedo*, which is the first dialogue where Plato begins to articulate his idealist theory of forms, he invokes a strong dualism between the soul and the body to assert the immortality of the soul.[5] Doing so, he depicts the body as being of a disruptive, polluting and evil nature which causes greed and war and obstructs the pursuit of knowledge.

When discussing 'The philosopher's detachment from the body' (64c–67b),[6] Plato argues that devotion to food, drink, sex and other bodily pleasures hinders the philosopher from seeking true knowledge (64d–e). To seek true knowledge, he insists, philosophers must rid themselves of their eyes, ears and 'the body altogether', because 'when the body is the soul's partner it confuses the soul and prevents it from coming to possess truth and intelligence [and] ... grasp that which really is' (66).[7] What really is, Plato claims (78b–80c), is a thing like the soul, which 'is immortal, indestructible, of a single form, accessible to thought, ever constant and ... true to itself' (80b).[8] Although Plato recognizes that even the living philosopher is embodied, he insists that as long as we are alive 'we shall be closest to knowledge if we refrain as much as possible from association with the body and do not join with it more than we must' (67a; see also Protevi, 2001, p. 125). If we use the body and its senses to investigate things, our souls will wander off in a 'dizzy drunken confusion', and only be capable of 'apprehending confused objects' (79c). Assuming that the fleeting, unstable and untrustworthy nature of the body and its perceptions constitute a fundamental constraint on being, on the soul and on our capacity to reach true knowledge, he further argues that the soul must seek to master and control the body, making it a 'slave and subject' of its rule (80).

The soul/body dualism and the soul's mastery of the body is reiterated in parts of *The Republic*. Here, Plato considers at length what training future philosophers need to undergo in order to nurture their souls and master their bodies. When discussing physical education, Plato argues that 'physical excellence does not in itself produce a good mind and character' (403d). Hence, the physical education of the philosopher in training should be 'simple and flexible' (404b). Moreover, the philosopher must learn to detest idleness, avoid filling his 'bod[y] with gases and fluids' (drunkenness has already been forbidden [398e, 403e]), and learn to take care of their own body and health so that intervention from medical doctors is only needed in case of 'injury or regular disease' (405d). While failure to do so suggests that one has a weak soul incapable of mastering the body, the ability to do so means that the pursuit of higher tasks avoids being disrupted by simple bodily matters (see also Protevi, 2001, p. 126). Thus, full attention can be devoted to the study of mathematics and pure philosophy, which Plato regards as the key subjects of philosophical education. The true philosopher, Plato insists, is someone whose 'pleasure will be in things purely of the mind' because 'physical pleasures will pass him by' (485d–e).

The soul's mastery of the body is further expressed through the isomorphic analogy which underpins Plato's theory of the state and renders the individual, society and the cosmos as being of the same form. Like the healthy body, which is mastered by the soul, the ideal, civilized and just society of the city-state that Plato outlines is governed by a philosopher ruler or a ruling class of philosophers. In the preliminaries, Plato refers to the civilized society as 'the true' society, 'like a man in health' (372e), which can only produce justice through a naturally given division of labour whereby 'a man specializes appropriately on a single job for which he is naturally fitted, and neglects all others' (370c). When he later considers justice in the state and the individual, Plato argues that there is 'no difference between a just man and just city' (435b) because they share 'the same three elements in the[ir] personality' (441c). Hence, the ruling class of philosophers must apply their reason and courage to manage the courageous spirits and physical appetites of the enterprising class of proprietors and businessmen (442; see also 441).[9]

Aristotle and the soul's mastery of the body

Plato's most famous pupil Aristotle arrived at Plato's academy with a long-standing interest in biology and medicine, and came to disagree in several respects with his teacher. Although it is not established that Aristotle ever outlined a systematic critique of Plato's idealist theory of forms, he provided a number of arguments that challenged it and the one-sided notion of the body which it implied.[10] Against Plato, Aristotle developed a 'hylomorphic' theory of substance which preoccupied him in several of his works and conceptualized form and matter as joined rather than separate.[11]

In his treatise *On the Soul*, Aristotle rejected Plato's separation of soul and body, emphasizing that 'the soul cannot be without a body ... [but] is *in* a body, and a body of a definite kind' (414a4–414a28). In other works he acknowledged that the generation of knowledge involves a combination of the soul's capacity for reasoning and rationality with the perceptual capacities of our bodily senses: the *Posterior Analytics* (100a3–100a9) argues that scientific knowledge involves moving from perception to memory, and from memory to experience; *Sense and Sensibilia* avers that the senses enable the soul to generate speculative and practical knowledge of things by feeding it with 'distinctive qualities of things' (436b18–437a3).

At the same time, Aristotle stressed that the soul and the body are distinct, and that they play a rather different role in the generation of knowledge. In *On the Generation of Animals*, he claimed that 'bodily activity is in no way associated with the activity of understanding' (736b28–29). In *On the Soul*, he insisted that 'the soul cannot *be* a body' (414a4–414a28), maintaining that the unity between soul and body was achieved as the soul mastered and organized the body (415b18; see also Protevi, 2001, p. 156).

Similar ideas are articulated in the *Politics*, where Aristotle portrays a city-state which, unlike Plato's republic, is no utopia but an already existing and very real entity populated by political animals (1254a). Here, Aristotle argues that the body and desire are naturally subject to the rule of the soul and the understanding. Even though this did not unequivocally imply the mastery of passive and unfree subjects,[12] Aristotle depicts a natural and beneficial (*sic*) division of labour between men and women, rulers and slaves, which reduces women and slaves to passive bodies: whereas 'something [that] is capable of rational foresight ... is a natural ruler and master, ... whatever can use its body to labor is ruled and is a natural slave' (1252a30).

Descartes' cogito, the autonomous soul and the passive body

While Aristotle remains one of the most significant thinkers to date, his proto-empiricist concern with sensual perception as a source of knowledge was fundamentally challenged as Descartes sought to found modern philosophy on rationalist principles. Aristotle had implied that knowledge of the world involved the knower turning one's senses outwards to objects in the external world. In contrast, Descartes came to assume that the senses were fundamentally unreliable and that knowledge had to start with the subject considering itself, indeed reflecting on itself, thus recognizing what is immediately and intuitively true.

In his *Discourse on Method*, Descartes (2007 [1637]) started to question the widespread assumption in ancient and medieval philosophy that humans could know the world and that such knowledge could exist in the first place. Since the truthfulness of what we perceive is doubtful, knowledge for Descartes began with doubt. By making the soul equal to thought and consciousness, he concluded that the only fact which we cannot doubt is the existence of thought and the thinking subject: the fact that I think proves that I have a soul, which makes me conscious that I exist. It is this reflexive exercise which enables Descartes to arrive at his famous *cogito, ergo sum*: 'I am thinking, therefore I exist' (in Part 4, p. 15; see also Cottingham, 1996, p. xxix).[13]

By establishing the soul's and the thinking subject's autonomy, it is this *cogito* which makes the soul independent of the body and clears the way for the Cartesian mind/body dichotomy. In the *Meditations on First Philosophy*, Descartes posits that the human, like the world, consists of two substances: *res cogitans*, which is the soul's thinking substance, and *res extensa*, which is the body's corporeal substance. Whereas the body's corporeal and extensive nature makes it dividable into different organs, Descartes states that he cannot imagine himself having 'half of a mind' (1996 [1641], p. 9). When he contemplates his soul and himself as a conscious being, he cannot differentiate any parts within himself (p. 59) because the soul's capacity for volition, understanding and perception cannot be identified as different parts of the soul; it is the same soul which wills, understands and perceives. In other words, he cannot imagine having half a soul because it is only as a whole that the soul is capable of thinking and imagining anything to begin with.

This is important because the soul's undividable nature means that it can be separated from the body and exist independently of what happens to the body. While the soul is united with the whole body (through the pineal gland, which he claims is the soul's seat in the brain), Descartes argues that the soul remains intact even if the body loses an organ such as an arm or a leg (p. 59). The soul's undividable and autonomous nature means that it can influence the entire body, but the body's dividable nature means that all its organs can have no impact on the soul; the soul can only be affected indirectly by the body, through the brain. This does not mean that we should exaggerate the soul's influence over the body in Descartes' philosophy;[14] his separation of soul and body does not imply that the soul can literally control or move the body.[15] However, it does enable him to emphasize the soul's capacity for thinking, reason and rationality and the body's mechanistic, passive and unreliable character.

Kant's underappreciated union of mind and body

The Cartesian *cogito* precedes a series of later contributions in the history of modern philosophy and western thought, perhaps most famously Kant's. It could be argued that the elevated status that Descartes' *cogito* gives to the soul and to the human rational subject prepared for Kant's transcendental idealism and his inverse Copernican turn, which renders our knowledge of the world dependent on our perspective on the world.[16] As we shall see, this is significant for how Kant views the relationship between the soul and the body, though his method and his conclusions are fundamentally different from those of Descartes.

In the *Critique of Pure Reason*, Kant (1998 [1770–87]) seeks to establish the limits of philosophy, science and knowledge. Since we cannot gain direct access to the world, Kant argues, we cannot know things in their own right, as *noumena*; we can only know things as they appear to us, as *phenomena*. In the preface to his *Critique*, he maintains that 'our rational cognition a priori ... reaches appearances only, leaving the thing in itself as something actual for itself but uncognized by us' (Bxx). Consequently, 'our representation of things as they are given to us does not conform to these things as they are in themselves but rather that these objects as appearances conform to our way of representing' (Bxx).

This view is reiterated in the 'Transcendental Aesthetic', where Kant details how it is only from 'the human standpoint' (A26/B42) that we can understand how things extend in space, and that it is only in a 'subjective' (A32–B53) manner that we can understand how things occur and endure in time. Further on, when Kant in the 'Transcendental Logic' develops his notion of the transcendental categories, he argues that it is these categories (of quantity, quality, modality and relation) which shape our understanding of the world and things in the world. On Kant, then, it is not the objects of our experience that give rise to these categories. Rather, it is the categories that make it possible to represent all the objects of our experience. In other words, Kant suggests that the world is not already given to us through our sensory perceptions of it or through qualities inherent in things themselves but that the mind orders the world by ordering how it appears to us.

For the sake of rhetoric effect, it might have been tempting to take this to suggest that Kant elevates the mind to such heights that he renders our bodily capacities of sensation and perception completely subservient in the generation of knowledge. However, the relatively marginalized status that the body and embodiment continue to endure in the social sciences, particularly in the literature on method and methodology, is by no means the result of a linear and univocal development in the history of western intellectual thought. Although certain commentators have implied a mind/body dualism in what they consider Kant's rationalist moral philosophy (e.g. Lakoff and Johnson, 1999; Hengehold, 2007), there has also been a number of efforts to unpack embodied aspects in Kant's philosophy (see, e.g., Shell, 1995; Svare, 2006; Palmquist, 2016). While we shall not examine these contributions in any detail here, they accentuate the value of teasing out some of the key ways in which Kant's understanding of the mind and its relationship to the body differs from that of Descartes.

Kant was dissatisfied with Descartes' rationalism, and the *Critique of Pure Reason* was an attempt to overcome the limitations of rationalism as well as empiricism. As pointed out by leading commentators, Kant develops a philosophy where synthetic knowledge is the combined product of our sensory perceptions of external objects and our cognitive capacity to order those perceptions (see, e.g., Guyer and Wood, 1998, p. 6). But Kant also takes issue with specific elements in Descartes' philosophy. In the 'Transcendental Analytic' he explicitly refutes Descartes' idealist view that the doubtful status of our perceptions means that we cannot have any knowledge of things that exist outside of us. Further on, in the 'Appendix to the Transcendental Dialectic' he rejects the Cartesian notion that the soul is a substance which can exist independently of 'corporeal nature'. As Kant emphasizes, that 'would take away ... all predicates of any possible experience' (A684/B712).[17] In place of a mind/body dualism, then, it could be argued that Kant formulates a notion of human nature which unifies the mind and the body (see, e.g., Palmquist, 1993, 2016; Guyer and Wood, 1998).[18]

Durkheim's society of minds and representations

Kant's mind–body synthesis did not prevent some of the pioneers of modern social science from reading him in a cognitivist and disembodied manner. Emile Durkheim, who is widely regarded as the founder of modern sociology, took significant inspiration from the transcendental idealism articulated through Kant's Copernican turn. As Lukes (1982) argues in the introduction to his now classic edition of Durkheim's writings on method, Durkheim was positive that sociology 'contributed to a renewal of philosophical questions' (p. 10) by 'offering sociological answers to fundamental Kantian questions' (p. 11).

This belief is perhaps most clearly articulated in a text where Durkheim (1982 [1909]) discusses 'The contribution of sociology to psychology and philosophy'. Here, Durkheim emphasizes that 'collective consciousness' rather than individual consciousness has to be the chief focus of sociology because 'It is society that informs our mind and wills' (p. 237). According to Durkheim, such influencing is achieved through a 'totality ... of human representations', which, in modern society, consists of the social institutions of 'religion, science, language and morality etc.' (p. 238). Hence, Durkheim seems to have been confident that sociology is able to understand the Kantian categories that shape how we represent, experience and understand the world, and how these categories emerge.

However, Durkheim pursued an explicitly rationalist agenda, which ignored the complexity in Kant's synthetic philosophy and instead embraced 'Descartes' method of doubt' (1982 [1895], p. 72) and 'cult of distinct ideas which [he viewed as] ... the basis of all science' (1973, p. 22).[19] In *The Rules of Sociological Method*, Durkheim (1982 [1895]) emphasized that sociology must pursue the rationalist study of social facts.[20] According to Durkheim, 'social life was made up entirely of representations' (p. 34), of 'collective believes, tendencies and practices' (p. 54), which should be investigated through observation and experimentation, by 'assum[ing] the state of mind of physicists, chemists and physiologists', and by 'feel[ing] [one]self in the presence of facts' (p. 37). It is this, Durkheim maintains, which enables sociology to grasp the *sui generis*, or unique character, of social reality, and establish itself as a distinct academic discipline independently from yet on par with the natural sciences.

So, even though Durkheim prescribed that sociology should utilize the kind of methods offered by the natural sciences, it was crucial for the existence and growth of sociology that it pursued its own research object.[21] Promoting an epistemic division of labour between the social sciences and the natural sciences, Durkheim stressed that 'there is a world of difference' between the physical and the moral environment (p. 44).[22]

As a consequence, the body and other phenomena which were associated with the natural world were excluded from Durkheim's research agenda for modern sociology and social science. Durkheim was particularly concerned that sociology should not be confused with psychology and biology. While acknowledging that a moral code or social norm may be felt within the person, Durkheim stressed that it is 'received through education' (p. 50). This, he asserted, is even the case with bodily practices such as habits of eating, drinking and sleeping (p. 54) and what he called 'mixed phenomena'. For Durkheim, mixed phenomena such as birth rates were of interest to the sociologist only in the sense that they were dependent on social institutions such as marriage laws (p. 55). Similarly, collective outbursts of emotion at social gatherings were viewed relevant only insofar as they were seen as the outcome of a shared existence and consciousness (p. 56).[23] Bodily practices as well as felt, embodied experiences were deemed irrelevant.

Durkheim's Cartesian tendency to disembody the study of sociology was further articulated in his late text 'The dualism of human nature and its social conditions' (1973 [1914]). Although Durkheim here makes a number of explicit references to Kant and none to Descartes, he argues that the contradictory forces of the soul and the body turn the human into a 'homo duplex'. The soul/body dualism is not an illusion, he points out, but confirmed by psychology, that is, by a universal feeling that '[human] nature is dual' (p. 151).

Furthermore, and with reference to his own study of the sociology of religious life, he associates the soul with the sacred and universal realm of moral conduct whilst locating the body in the profane and individual realm of sensation. Whereas morality involves the pursuit of impersonal goals and a disinterested 'attachment to something other than ourselves' (p. 151), 'Our sensory appetites are ... egoistic' because it is only ourselves we satisfy by satisfying our hunger, thirst and other bodily needs. With a nod to Kant's moral philosophy of duty, Durkheim explains that the pursuit of 'moral ends ... caus[es] a split within ourselves' because it necessarily means that we must sacrifice the satisfaction of our own egoistic appetites (p. 152).[24]

At the same time, Durkheim does acknowledge that our sensory experiences cannot be mastered, grasped and made fully intelligible through concepts. Thus, 'we cannot understand things without renouncing a feeling for their life, and we cannot feel that life without partially renouncing the understanding of it' (p. 153). Towards the end of his career, then, Durkheim was certain that sociology had no business investigating the sensory realm of bodies. With regard to the soul's mastery of the body, the matter was different. Though recognizing that we cannot fully master our bodies, he insisted that we nevertheless must strive to master them and overcome the boundaries they impose on us: since 'society surpasses us, it obliges us to surpass ourselves ... by depart[ing] from ... [our] nature ... [and] do[ing] violence to certain of our strongest inclinations' (p. 163).

Now that a century and more has passed since Durkheim made these claims we are aware that sociology and social science research has developed into a more heterogeneous realm than envisaged by Durkheim. This is not least evident through the increasing attention that social scientists have started to show in the body and in issues of embodiment. Indeed, we would be hard pressed to find examples of Durkheim's rules of sociological method being applied even in the kind of disembodied research on the body which we discussed in the first part of this chapter.

Hence, we are not suggesting that there is a direct causal link between the disembodied legacy of Durkheim's sociology and the disembodied methodologies that have been assumed in certain parts of social science research on the body. Neither do we imply that there is a neat linear pattern running all the way from the disembodied sentiments that we have traced through classical and modern philosophy, via Durkheim, through to the contemporary examples of disembodied scholarship that we discussed above. However, by establishing the body as an unreliable source of knowledge, and by steering the social sciences away from embodied issues and methods, the disembodied legacy which these thinkers offer has raised the intellectual and institutional threshold for investigating the body in embodied ways.

DISEMBODIED ANSWERS TO BODILY QUESTIONS

In the next chapter we shall move on to consider methodology and scholarship that has sought to investigate bodies and embodiment in more embodied ways. But, to get a better sense of what this may involve, we must conclude this chapter by summarizing what answers the disembodied approaches we have examined here provide to some basic but fundamental questions about the body: that is, what answers do they provide to the *ontological* question of what the body is, to the *epistemological* question of what we can know about the body and to the *methodological* question of how we can go about knowing the body?

Disembodied ontologies

As our previous discussion suggests, a mind/body dualism runs through the ontologies of Plato, Aristotle, Descartes and Durkheim. Without wanting to reduce these formidable intellectual figures to one homogeneous position, it is worth noting that all four of them construct the body as fundamentally different from and subservient to the soul, mind and consciousness. Durkheim's references to Kant even seem to imply (though inappropriately so) that Kant assumed a mind/body dualism by being solely concerned with consciousness.

The mind/body dualism is particularly strong in Plato and Durkheim, both of whom stressed that the body is an entity driven by egoistic and hedonistic desire, and which therefore constitutes a threat to moral conduct and social cohesion unless it is managed and controlled by the soul. While such dualism is rarely if ever expressed in contemporary research on the body, the prevailing tendency to conceive of the body as an object of social regulation, discursive construction and conscious understanding rests on an implicit dualism. As a result, this literature effectively turns the body into a passive object of the mind's ability to know it.

Disembodied epistemologies

There is more variety between their epistemological positions. According to Plato, we can only know the idea of the body; that is, we can only have true knowledge of the body as an abstract ideal form. Thus, we cannot know but only have flawed beliefs and opinions of our own bodies. In contrast, Aristotle held that we can truly know the noumenal reality of the body, though this is limited to the natural and physical body which is the object of science, biology and medicine. In turn, Descartes, who concluded that we can only have certain knowledge about the soul, nevertheless deduced that the autonomous and undivided nature of the soul enables us to have rational and truthful knowledge about the mechanistic workings and divisive structure of the body.

As we highlighted above, Kant's inverse Copernican turn concluded the break with classical and medieval philosophy which was initiated by Descartes. Thus, all possibilities of achieving true knowledge about the world, about objects and about the body in itself were eliminated. Kant, then, allows us to 'only' know the phenomenal nature of the body; that is, the body as it is understood from the human perspective. About a century later, it seems that Durkheim may have appreciated Kant's Copernican turn more than Kant's effort to unite the mind and the body in a human subject. Durkheim, then, is confident that natural scientists – but not social scientists – can know the facts about the body.

Despite their differences, what seems to join these figures is the more or less implicit assumption that we cannot know how people, in everyday social life, experience their bodies, relate to their bodies, or express themselves through their bodies. Although some of the disembodied research on the body in contemporary social science has acknowledged the importance of investigating this lived reality of embodiment (see, e.g., Turner, 1984), its refraining from doing so echoes the epistemic limit implied by Durkheim as well as by Plato, Aristotle, Descartes and Kant. Focusing on knowing how the body is an object of social and discursive construction, it leaves little direction for getting to know the lives and bodies that are worked upon by social and discursive forces.

Disembodied methodologies

Unsurprisingly, the methodological answers offered by these disembodied approaches are shaped by their epistemic limits. Plato maintained that we can know the true nature of the body through ideal concepts that are innate yet realized and understood through careful philosophical training and dialogue. Aristotle challenged Plato's idealism by arguing that we can know the noumenal reality of the body through a combination of sensory experience and philosophical concepts.

Although Descartes' interest in science and medicine was on a par with that of Aristotle, his rationalism has certain similarities with Plato's idealism, and his rationalist inclinations led him to stress that we can know the mechanistic nature of the body through clear and distinct ideas formed in the mind. In contrast, Kant's synthetic philosophy has certain similarities with Aristotle. Even though Kant's Copernican turn meant that we cannot know the body as a thing in itself, the phenomenal body that we can understand from a human perspective can be known through a combination of sensory experience and concepts which are ordered according to the transcendental categories of the understanding.

As we mentioned above, Durkheim made a number of references to both Descartes and Kant, and in some way he combines the two as he implies that natural scientists can know the facts about the body through scientific observation and experimentation and through the development of distinct Cartesian concepts. But even though the body's capacity for sensation plays an important methodological role not just in Aristotle and Kant but also in Durkheim's emphasis on visual perception, it is made subservient to the mind's capacity to order, rationalize and understand sensory input according to concepts; in itself, Durkheim insists, bodily sensation, perception and experience is a too fleeting and unreliable source for generating truthful knowledge.

We do not deny the importance that conceptual reasoning plays in generating, and sharing, knowledge. As social scientists, it is important that we try and articulate appropriate concepts for understanding our embodied experiences and bodily expressions, and the embodied ways in which we interact with others. Such concepts can certainly help us talk about and write about this lived embodied reality in ways that may be meaningful to others. The trouble with the disembodied heritage examined above is that its methodologies offer no means for exploring the lived embodied reality which may inform such concepts.

This is also a key problem with contemporary research on the body which assumes a disembodied approach. By focusing on the ways in which bodies are subjected to social and discursive practices, much can be known about the norms, institutions and texts that seek to shape and understand our bodies. However, it does not allow us to know much about the lives and bodies that are subject to these practices, or how we, with our bodies, cope with, manage, or transgress them.

Thankfully, this is not all that is on offer. As we shall see in the next chapter, there have been a number of efforts across the social sciences to approach the body in embodied ways, and to work with the body as a subject and object of knowledge. These developments provide a crucial platform in our further attempt to embody the methods and methodologies of social science scholarship.

Questions and exercises

1. In what ways may some of the social science research which explicitly deals with the body be regarded as disembodied?
2. How may this tendency be traced back through the history of western thought?
3. What are the main differences between Plato's and Aristotle's view of the body's relation to the soul?
4. Read through the notes relating to the section on Descartes. On what basis is it possible to suggest that Descartes himself was not a Cartesian?
5. Assume a Kantian position: In what sense can the body be known and understood?
6. Why did Durkheim exclude the body from the research agenda that he developed for modern sociology?
7. What might be missed by investigating the body in disembodied ways?

2 EMBODIED METHODOLOGY

MY BODY AND OTHER BODIES

In his ethnographic study of boxing, the sociologist Loïc Wacquant (2004) offers a number of passages describing his embodied experience of becoming an apprentice of the pugilistic art. Wacquant insists that the body is our primary tool for scholarly investigation and observation, and he spent three years among the amateur and professional boxers and trainers of the Woodlawn Boxing Gym in South Side Chicago to learn the art of boxing and to learn about the everyday struggles of his research participants.

We find Wacquant's writing particularly embodied towards the end of his book as he recollects the time around his first official match in the ring. Sharing some of the stress and anxiety he experienced the days leading up to the match, he fears that he has 'caught a cold' because his neck hurts, his nose runs and his 'throat is in knots', and admits trying to combat feelings of weakness and nervousness by taking a hot bath and 'gulping ... chocolate mints, ... cookies, and ... ice cream', which only adds to his 'malaise' (p. 244). Some pages later, Wacquant portrays the beginning of the fight by describing how 'Punches are raining from the get-go', how he 'take[s] a straight right smack in the face', and how 'punches hurt more – a lot more!' when thrown '[w]ith competition gloves ..., which are ... less padded' (pp. 250–1).

While these incomplete quotes cannot do justice to Wacquant's rich narrative, we are struck by the vibrant intensity with which Wacquant describes his own feelings. Elsewhere in the book, he offers detailed accounts of other Woodlawn boxers, such as Butch, who is Wacquant's first sparring partner. Wacquant (p. 72) describes what Butch is wearing ('a sleeveless blue jersey'), his warm-up punches (which 'ferocious[ly]' cut 'through the air'), his imposing figure (he is 'a head taller' and 'his torso and arms are like rolling balls of ebony and glistening under the gym's pallid ceiling lights'), and he questions his own decision to spar with this 'colossus' (who he 'suddenly find[s] ... even more muscular than usual'). Throughout the book, Wacquant's narrative includes several references to physical appearance, bodily actions and movements, visceral feelings and experiences.

Although Wacquant's study is a good example of embodied research, his expression of visceral feelings and lived experiences is not unique. In her book titled *The Vulnerable Observer: Anthropology That Breaks Your Heart*, Ruth Behar (1996) shares her personal experience of grief and loss. With 'intense regret and self-loathing' (p. 21) she mourns her grandfather, who she knew was dying of cancer in Florida when she decided to spend the summer doing fieldwork in Spain. The same morning that her grandfather dies, she wakes up, in Spain, 'ill and nauseous with pain in my bones' (p. 68). Flying back for the funeral, she feels 'angry and imposed upon that [she] can no longer speak of him in the present tense' (p. 75). For Behar, the 'body is a homeland ... where knowledge, memory and pain is stored' (p. 134).

This is not Wacquant's cup of tea. In his later appeal to a sociology of flesh and blood (Wacquant, 2015), he takes issue with Behar's approach, claiming that '"writing vulnerably" by injecting large doses of "subjectivity into ethnography" would cause us to drown in the bottomless whirlpool of subjectivism' (p. 5). Instead, he argues that we should dare to 'become "vulnerable observers" in our practice of fieldwork' by 'div[ing] into the stream of action to the greatest possible depth' (2015, pp. 4–5). We need to become more involved in the settings we study, he insists, and develop the 'spunk' and 'daring' that is required of us to do so. But rather than diving in blindly, we need to 'dive and swim along with method and purpose' (p. 5), and we need to return from those settings if we are to make sense of them and tell others about them (p. 7).

Given Wacquant's own immersion in the world of boxing it is difficult to ignore the masculine flair of his rhetoric. But beyond that, his appeal raises a number of questions which have a wider significance for how we might work with embodied research methods: How do we choose to approach the empirical settings we study? Which techniques and procedures do we make use of to generate empirical material from them? How do we go about interpreting and analysing our material? And how do we choose to write about and represent the insights we draw from it?

We shall briefly return to these questions towards the end of this chapter but deal with them in a more concrete manner as we discuss the various details of embodied research methods in Chapters 3 to 8. Before we do that we need to enquire further into the ontological and epistemological questions which are actualized by Wacquant's appeal. To get a sense of how one might go about investigating the body in an embodied manner, we need to have an understanding of what the body is and what we can know about it from an embodied perspective. To do so, it seems appropriate to begin with Merleau-Ponty's embodied phenomenology.

EMBODIED PHENOMENOLOGY

Maurice Merleau-Ponty is often regarded as the founder of embodied phenomenology and his work is widely used by social scientists and other scholars engaging in embodied research (including Loïc Wacquant). As we noted in our introductory chapter, Merleau-Ponty rejects the mind/body dualism, locates the mind in the body and underscores that the body is a subject which both thinks and perceives.

Perception, not reflection

In his main work *The Phenomenology of Perception*, Merleau-Ponty (1962 [1945]) elaborates his notion of the lived phenomenal body, opposing it to the mechanical body of physiology and empiricist psychology and setting it against the reflective body of transcendental philosophy and Gestalt psychology. He rejects Descartes' notion of the *cogito* and Kant's notion of the transcendental ego as fundamentally self-reflective moves which reduce the unreflective object of our thought to the conscious and reflective subject of our thinking (see e.g. pp. x, 71–2, 429). In a different vein, he calls empiricist psychology 'unsituated thought' which fails to read other people's feelings when observing them (p. 27) and which turns the experience of subjectivity into an object of scientific investigation that can only see a body as others see it (pp. 108–9).

Like all phenomenology after Kant, Merleau-Ponty insists that we can only know the world from the perspective of human understanding. But for Merleau-Ponty phenomenology is a matter of re-achieving direct and primitive contact with the world, and this is something that can only be achieved through the deeply embodied process of perception, not through reflection. Whereas perception and sense experience is a pre-objective realm (p. 14) and the very foundation of phenomenal existence which enables our vital communication with the world (pp. 11, 61), 'reflection obscures what was clear' (p. 11).

The centrality of the body to this project is reinforced through numerous claims throughout the book: Merleau-Ponty argues that his body is his 'point of view upon the world' (p. 81), the basis of consciousness, and 'the vehicle of being in the world' (p. 94); he further maintains that 'the body expresses existence at every moment' and that 'existence realizes itself in the body' (p. 192); indeed, the body is the knowledge-acquiring apparatus which enables us to conceive our own existence (p. 409).

It is the body, then, and the perceptual nature of the body, which involves us in things (p. 215) and puts us in contact with the world, with other people and with ourselves. According to Merleau-Ponty, we understand others through our bodies because we perceive others through our bodies (p. 216). Unlike Kant, who assumes that our understanding of the world is shaped and given by innate categories or schema in the mind, Merleau-Ponty argues that our schema of understanding are body-schema, which emerge through our perceptual engagement and intentional coping with the world (see also Taylor, 2003). Hence, the body is not only a closing off from the world but an opening up (p. 191).

The social flesh

The profoundly social nature of our bodies is further elaborated through his concept of 'the intentional arc': while our body schemas make us 'totally' aware of our own posture in the intersensory world (p. 114), the dynamic form of the intentional arc subtends in a broader sense our consciousness, desire and perception by projecting around us our past and future, our human setting, and our physical, ideological and moral situation (p. 157).

In Merleau-Ponty's (1968 [1964]) unfinished work *The Visible and the Invisible*, the perceptual, social and intersensorial dimensions of the body are given a stronger ontological meaning through his metaphorical concept of 'flesh'. 'Flesh' is the united fabric which joins the flesh of the body with the flesh of the world and embeds our bodies in the world by connecting the subjective and objective dimensions of the body as perceiving and perceived. His choice of the word 'flesh' is quite significant, because he inverts Saint Paul's and Augustine's pejorative concept by accentuating how the body's sensible and sensing flesh enables it 'to grasp and commune with the world of sensible things' (Shusterman, 2005, p. 155).

Learning and knowledge

However, Merleau-Ponty's concepts of body schema and the intentional arc even play a more specific role in the *Phenomenology* as he seeks to establish the body as an active and intentional medium of learning, habit, skill and knowledge. It is the body that 'causes' things to exist for us intentionally (1962, p. 169) by generating body schemas that involve my body in the world and anchoring it in objects that face it with tasks (p. 115). And as lived experience tends towards our habits and bodily functions (p. 158), it triggers bodily effort which over time turns into familiar habits that open us on to the world and correct how we engage with it, whether we are reading, typing or playing a musical instrument (p. 167).[1]

Learning, then, means recasting our bodily schema (p. 177) and changing how we engage with the world and the objects we encounter in the world. Further, the intentional arc projects us and situates us in the environment so as to shape our perception of the world and how we purposefully, practically and skilfully respond to it and cope with it (p. 157). Indeed, the embodied nature of human existence is so fundamental that, for Merleau-Ponty, it may not be necessary to think but sufficient to feel a shared project (p. 517). Language too is embodied. Words have a physiognomy which we try to grasp in their living expression. Reading the word 'warm', he says, makes one prepare to feel warm, and as 'the word pushes its way through my body ... I have the feeling ... of a ... numbed fullness which invades my body, and which ... imparts to my mouth cavity a spherical space' (p. 275).

Observation, not introspection

Now, for the social scientist eager to create lively accounts of how people express themselves through their bodies and how we experience social life with our bodies, we must ask ourselves how we can access these kinds of embodied feelings, this lived embodiment or *corpse propre* that Merleau-Ponty speaks about. Defining feelings and emotions as 'solipsistic' data of introspection (pp. 28, 417), he rejects introspection as an insufficient source of knowledge (p. 110). Moreover, he discards reflection as a way of accessing the inner experience of our bodies because intrinsic characteristics are expressed in appearance (p. 28). Yet, we cannot observe our own body because observation turns our body into something outside ourselves (pp. 106–7).[2] Furthermore, the immediacy of perceptual experience means that my reflection on that experience is always post-hoc (p. 276).[3] Whereas this leads Merleau-Ponty to argue that we gain knowledge of others and their feelings through our relations with them – that is, by perceiving and observing their behaviour (pp. 110, 414) – he regards any attempt to try and engage directly with lived inner experience pointless.[4]

Although much of the secondary literature on Merleau-Ponty in social and organization studies (e.g. Hindmarsh and Pilnick, 2007; Yakhlef, 2010) seems to accept his turn away from the lived inner experience of felt emotions and affects, a number of prominent contributions provide reflective accounts of felt experiences in writings where they also make use of Merleau-Ponty's work (see, e.g., Charlesworth, 1999; Wacquant, 2004). Before affirming this take, however, we need to discuss the epistemological challenges of engaging with lived inner experience in this way, and Merleau-Ponty's reasons for turning away from it.

Phenomenological limits, pragmatist openings and personal reasons

The pragmatist philosopher Richard Shusterman (2005) offers quite a unique treatment of these issues. According to Shusterman, Merleau-Ponty 'celebrat[es] the body's role in expression', but does not 'listen to what the body seems to say about itself' (p. 151). For Merleau-Ponty, conscious somatic sensations of how we move, position and use our bodies are not lived experience as such, but merely thematized 'representations of bodily experience' (p. 152). Hence, Merleau-Ponty (1964 [1960], p. 97) insists that our body guides us only when tacit but not when we analyse it or when we self-consciously focus on what we feel within it rather than with it (Shusterman, 2005, p. 153).

Unhappy with Merleau-Ponty's silencing of bodily experience, Shusterman offers a systematic argument for why we should end this silence. Central to his argument is a distinction he makes between Merleau-Ponty's notion of primary consciousness, and what Shusterman calls conscious body sensations. Whereas primary consciousness concerns the fact that my body is perceived by me and includes feelings such as hunger, pain and pleasure, conscious body sensations are 'not ... opposed to thought but' inclusive of 'conscious, experiential body-focused thoughts and representations' (2005, p. 156) which enable me to develop my somatic awareness.[5] Examples may include dancers glancing at their feet to ensure that they are 'properly turned out' (p. 156) and runners making sure that they keep their head up even when going up-hill. According to Shusterman, external sensations such as seeing and hearing, and kinaesthetic feelings gained from bodily movement and orientation, aid performance because they tell us what condition our bodies are in (p. 156).[6]

Further on, Shusterman disputes Merleau-Ponty's assumption that spontaneous intentionality can explain all the marvellous things that our bodies can do. Claiming support from William James' emphasis on introspection and John Dewey's stress on bodily consciousness, he points out that there is rehearsed intentionality too, particularly when we learn to do new things with our bodies, whether reading and typing, cycling, swimming or skiing. Rather than erasing unreflective behaviour, the focusing and reflection involved in somatic awareness complements our somatic spontaneity and helps us observe our bodies. It does not remove us from the outer world of action but may improve performance by increasing our experience, correcting bad habits and preventing breakdown of spontaneous competencies.

In summary, Shusterman concurs with Merleau-Ponty that full transparency of our bodily feelings is impossible. However, he rejects Merleau-Ponty's dichotomy between lived experience and representation, pointing out that bodily self-observation and reflection are part of our everyday practice and ordinary experience (p. 175). When Merleau-Ponty discards the possibility of reflecting about our immediate bodily experiences, Shusterman argues that he ends up with a purely intellectualist way of working over the body (p. 176) which privileges the logical concerns of philosophy to such an extent that it fails to make common sense. While Merleau-Ponty's embodied phenomenology offers a rigorous philosophical treatment of what the lived body is and what we can know about it, Shusterman's 'somatic pragmatism' opens a possible way out of the epistemological limits that prevent us from exploring inner feelings and experiences. Thus, it promises to take us beyond Merleau-Ponty's idealization and silencing of lived embodied experience, a silencing which paradoxically reinforces the mind/body dualism that he set out to subvert (see also Levin, 1985).

Ironically, there may have been personal and embodied reasons for Merleau-Ponty's philosophical decision. Shusterman (2005, p. 177) suggests that Merleau-Ponty's resistance to explicating somatic feelings does not merely emerge out of his phenomenological philosophy but is related to his childhood. In personal conversation with Jean-Paul Sartre, Merleau-Ponty had allegedly indicated that he struggled through philosophy to reconquer the spontaneity and happiness of a lost childhood. While his father died when Maurice was 10 years old, his loving mother was his rock. Sartre recalls this beautifully:

> Everything had been too wonderful, too soon. The form of Nature which first enveloped him was the Mother Goddess, his own mother, whose eyes made him see what he saw. ... By her and through her, he lived this 'intersubjectivity of immanence' which he has often described and which causes us to discover our 'spontaneity' through another. (Sartre, in Shusterman, 2005, pp. 177–8)

As Shusterman explains, Sartre suggests that Merleau-Ponty embarked on his philosophical project in order to reconquer the spontaneity and harmony he had experienced during his childhood, particularly through the close relationship with his mother. While his mother was alive this hopeful connection stayed alive, but when she died it disappeared: 'Merleau-Ponty was devastated and essentially "became a recluse"' (ibid, p. 178).

EMBODIED DIFFERENCES AND INEQUALITIES

There is at least one more problem with Merleau-Ponty's embodied phenomenology which Shusterman does not resolve. Both Merleau-Ponty and Shusterman are largely preoccupied with embodied performance and the improvement of 'normally' functioning bodies. They seem unconcerned with how different our bodies are. Assuming a universalist perspective on the human body, neither of them pays any attention to embodied disabilities and impairments, nor do they show any concern with the different ways people embody gender, sexuality, race, class and age. At the same time, a number of disability scholars, feminist scholars and critical scholars of sexuality, race, class and age have engaged with Merleau-Ponty's work in order to explore the lived embodied reality of these issues (see e.g. Charlesworth, 1999; Allen, 2004; Young, 1980; Ahmed, 2006, 2007; Antoninetti and Garrett, 2012).[7]

Since we are more concerned with advancing embodied methodology in general than with Merleau-Ponty's embodied phenomenology as such, we shall not investigate how these contributions relate to his work. Instead, let us move on to consider some examples of how embodied differences have been brought to life in written research accounts, with or without Merleau-Ponty.

Throwing like a girl?

An appropriate place to start is Iris Marion Young's (1980) classic piece 'Throwing Like a Girl'. Here, Young shows how the embodied nature of gender is imbricated with processes of socialization which influence what girls and boys can do, beyond but not regardless of gendered physical differences. Her starting point is a text on 'The Upright Posture' and the use of lateral space by the phenomenologist philosopher Erwin Straus (1966), where Straus comments on a study of 5-year-old girls and boys faced with the task of throwing a ball. As Young shows, Straus states that the girl in the photographs, unlike the boy, fails to throw the ball with any force because she does not involve her entire body in the act of throwing:

> The girl ... does not stretch her arm sideward; she does not twist her trunk; she does not move her legs, which remain side by side. All she does in preparation for throwing is to lift her right arm forward to the horizontal and to bend the forearm backward in a pronate position. (Straus, in Young, 1980, p. 137)

While Straus is puzzled by this difference and fails to explain it by anything other than what he calls a mysterious 'feminine essence', Young affirms that girls and boys, women and men, have a tendency to approach physical tasks in quite a different manner. With reference to women's involvement in sports and manual tasks, she argues that 'Women often approach a physical engagement with things with timidity, uncertainty, and hesitancy' (Young, 1980, p. 143). Like Straus, she maintains that women would not put the whole body 'into fluid and directed motion' but concentrate the motion 'in one body part' and not 'reach, extend, lean, stretch, and follow through in the direction of her intention' (p. 143). The consequence is, Young argues, that women's lived space is far more limited than men's space.

However, Young offers a rather different explanation from Straus, which alerts us to the limitations of malestream phenomenology. By complementing insights from Merleau-Ponty's (1962) notion of 'intentional movement' with de Beauvoir's (2011 [1949]) notion of 'women's situation', Young emphasizes how the bodily practices available to women are shaped and constrained not merely by physical constitution and muscular strength but by the expectations and norms that women encounter in a patriarchal and sexist society. This goes deeply back to the sexually differentiated upbringing of girls and boys, which at least was dominant in the US context that Young worked in at the time. Referring to a study of girls' and boys' play in school and after-school activities, Young states that 'girls are not often asked to perform tasks demanding physical effort and strength', but boys are 'more and more' as they 'grow older' (1980, p. 153).

Reflecting back some 25 years, Young (2005) admits that much has changed since she first wrote the piece: girls are now more likely to be encouraged to engage in sports and other physical activities where they might get injured or tear their clothes. Moreover, as we showed in our introductory chapter, many women, men and trans people challenge the constraints of stereotypical gender roles in a variety of ways, through a diversity of bodily actions and practices (see, e.g., Halberstam, 1998; Parkins, 2000; Na and Choo, 2011; Reestorff, 2014; Thanem and Wallenberg, 2016; see also Brewis and Sinclair, 2000). What gives Young's text its enduring legacy is its sharp analysis of how gendered differences and inequalities are inscribed through processes of socialization; although the text lacks vivid descriptions of the physical acts she writes about, it shows in a most tangible way how social context and conditioning may affect embodied action.

Protesting like a girl

Wendy Parkins' (2000) text 'Protesting Like a Girl' provides a necessary transgression of Young's discussion. As we noted in our introductory chapter, Parkins draws attention to the physical and psychological toughness that was embodied by suffragette activists, more than half a century prior to the embodied situatedness of women that Young writes about.

Focusing on the working-class suffragette Mary Leigh, Parkins gives lucid insights into these dimensions by utilizing material from historical archives and UK newspapers at the time. In particular, she shows the courage and steadfastness that was required to display suffragette colours in public at the risk of physical attack from political opponents and the police; the tenacity demonstrated by suffragettes who went on hunger strike during imprisonment; and the strength and agility exercised when climbing onto rooftops to protest against anti-suffragette meetings. Parkins' quote from *The Liverpool Courier* gives a lively example of one such incident:

> Window-panes fell one after another before the unerring shots of the attacking force. … [The women] devoted themselves to handing up bricks and stones to their comrade on the roof [Leigh], and she transferred them through the windows of the Sun Hall with a dexterity which was nothing short of marvellous. … The officers commented upon the remarkable agility and daring which she must have displayed to reach this extraordinary position for attack. (*The Liverpool Courier*, in Parkins, 2000, p. 66)

As Parkins notes, Leigh 'did *not* throw like a girl' (p. 66; original emphasis). However, Leigh was very quiet about her own achievements, and despite her likely knack for public speaking little is known about her own experience of fronting suffragette politics. One exception is the following quote extracted from Parkins, where Leigh, in a statement to her solicitor, gives a detailed though emotionally tempered account of the appalling treatment she was subjected to in prison while going on hunger strike:

> Sunday noon, four wardresses and two doctors entered my cell and forcibly fed me by the tube through the nostrils with milk. ... I remained in the padded cell until Monday evening, October 4. Since then I have been fed through the nostril twice a day. The sensation is most painful – the drums of the ears seem to be bursting and there is a horrible pain in the throat and breast. The tube is pushed down twenty inches. I have to lie on the bed, pinned down by wardresses, one doctor stands up on a chair, holding the funnel end at arm's length, so as to have the funnel end above the level, and then the other doctor, who is behind, forces the other end up the nostrils. (Leigh, in Parkins, 2000, p. 71)

Parkins, then, unfolds a rather different body-subject from Merleau-Ponty's phenomenology, and the historical quotes in her account add a vivacity which reinforces the notion of a politics which is actively embodied yet victim of violence and abuse.

Experiencing racism

The felt and enfleshed experiences of embodied difference are perhaps more directly expressed in Chicana feminism, which remains a unique force in understanding and combating the mechanisms of racism and sexism that raced women are subjected to in western patriarchal societies. Personal, embodied experience plays an important role in Chicana scholarship. In the now classic anthology *This Bridge Called My Back*, Naomi Littlebear (1983) offers a particularly strong narrative of her embodied memory of racist and sexist violence. Here, Littlebear depicts how she and her cousin escaped, horrified and 'on rubber legs', from a group of violent leather jacket boys 'screaming' and 'thrashing' at them with 'harsh stiff leather' and 'metal teeth zippers battering ... [their] bewildered bodies' (p. 16).

Victimhood, however, is not the end point of Chicana feminism. Refusing to be dragged down by their suffering of racism and sexism, Chicana feminists transform their pains into political energy in the struggle for a free and equal society. Stating that she is 'tired of making a tragedy of our lives' (p. iv), Anzaldúa, in her foreword to the second edition of the anthology, emphasizes the connectedness of this struggle, as it joins 'white black straight queer female male' people (p. iv). Further on, Moraga stresses the 'physical and psychical' nature of the fight for freedom, which 'is about intimacy, a desire for life between us, not settling for less than freedom even in the most private aspects of our lives' (p. xix).

While the lived engagement of Chicana feminism is not frequently referenced by social scientists, the transformation of painful experiences into political action is an important idea which has been reiterated more recently in other parts of feminist research (see, e.g., Ahmed, 2004, 2017; Probyn, 2004). We shall not discuss this in more detail here but return to it in Chapter 9 where we expand upon the political role of embodied research. It is striking, however, that there is far more scholarship by women than by men on issues of embodied difference. Few male academics seem prepared or even bothered to engage with the felt experiences of social life, their own or others'.[8]

Embodying poverty

An exception to this tendency is Simon Charlesworth's (1999) book *A Phenomenology of Working Class Experience*. Here, Charlesworth draws attention to the embodied expression and experience of class by examining the personal consequences of poverty for young men and women living in his deprived home town of Rotherham in Northern England. Insisting that 'one ... has to be amongst this life ... and ... be sensitive to what it makes of them' to be able to trace and write their life stories (p. 2), Charlesworth illustrates his rigorous analysis of working-class experience with vibrant and intimate reports based on interviews and conversations.

Concerned with how his participants 'feel, and why; and what those feelings reveal about their relationship to social reality' (p. 8) he provides detailed accounts: of what it feels like to constantly fear losing one's job; of the guilt and boredom participants feel from being out of work; of the disgust they feel from being exploited in temporary underpaid jobs of hard physical labour (see, e.g., p. 176).

Charlesworth notes that bodily posture, body language and other embodied behaviours constitute an important part of how class is lived and experienced. However, he declines to include such descriptions of his participants, fearing that it would risk reinforcing disrespectful stereotypes of working-class people (p. 24). By connecting 'the realm of economic organization and the most personal, intimate realms of human life' (p. 81), he points out that his aim is to detail and analyse the lived effects of stigmatization, not to contribute to it. Other scholars may reason differently about this, and make different choices, but Charlesworth's decision enables him to read personal narratives through phenomenological concepts (especially from Merleau-Ponty and Bourdieu), and add analytical sharpness without violating the personal integrity of his participants or the fleshed style of their narratives.

EMBODIED ANSWERS TO BODILY QUESTIONS

With the risk of distancing embodied research from the vivid expressions and lived experiences that it thrives on, we need to close this chapter too with a meta-theoretical discussion: What answers do the embodied approaches we have discussed above provide to the *ontological* question of what embodiment is? How do they respond to the *epistemological* question of what we can know about the lived reality of embodiment? And what answers do they offer to the *methodological* question of how we can go about investigating embodied lives? In other words, to what extent can we access embodiment as an object of knowledge, how can we use our own bodies as subjects of knowledge to do so, and what is this lived embodied reality which we strive to access?

Embodied ontology

For Merleau-Ponty 'embodiment' denotes the lived, sensorial, perceptual, experiential and active nature of the body, a body which incorporates the mind, and which is embedded and incorporated in the world. This is a pre-reflective body, which is uncorrupted and untainted by our reflective and representational judgement. It is a sensual and perceiving body, which constitutes our first and primary contact with ourselves and the outside world; it is that material fabric through which we experience and know the world and ourselves in the world; and it is the medium through which we make sense of, think about, talk about, and enact the world whilst being embedded in it and shaped by it. The embodiment emerging out of Merleau-Ponty's work is a lived body which harbours, processes and expresses feelings and emotions, language, gestures and movements, thinking and learning, actions and behaviour. In contrast to the disembodied methodology and approaches we discussed in the previous chapter, this embodiment connotes a body which is both subject and object.

As Merleau-Ponty seeks to explain how this embodied reality is lived by all of us, he unfolds it in a universalist manner which seems oblivious to the strikingly different ways in which our lives are embodied. However, this is fleshed out in a variety of research examining specific ways in which lived embodiment may be disabled, aged, gendered, raced or classed. The various examples we have discussed here provide little explicit commentary on the ontological status of lived embodiment as such, but they do show how differently people's lives are embodied and imbricated with the social and economic conditions they live under.

In feminist scholarship this is primarily a question of gender. As we saw above, Young's (1980) discussion of physical activities underscores how gendered inequalities and constraints are embodied yet inscribed through the different processes of socialization that girls and boys are traditionally subjected to. In contrast, Parkins' (2000) work on suffragette activism shows how the constraints of stereotypical gender roles are challenged and transgressed through embodied political action. In Chicana feminism embodiment assumes a lived reality where experiences of gender and race intersect. The lived reality they depict is one of pain and suffering, but also one which inspires and instigates political action (see especially Moraga and Anzaldúa, 1983).

Finally, one might think that class is what defines lived reality in Wacquant's (2004) work as well as in Charlesworth's (1999) work. After all, both their studies were conducted in working-class contexts, but this comes through more strongly in the latter than it does in the former. Where Charlesworth is concerned with the lived experience of unemployment, poverty and dirty work, Wacquant explores life around a boxing gym in a black working-class neighbourhood. Thus, Charlesworth presents us with a classed notion of lived embodiment surrounded by feelings of stigmatization, depression and anxiety expressed through postures and gestures adopted to ward off constant threats of violence and physical assault. Wacquant does not neglect these aspects of working-class poverty but portrays a lived reality more incorporated by the bodily habits, activities and skills which increasingly became part of his life as an apprentice boxer.

Embodied epistemology

Now, in what sense can we know this realm of embodied experiences and expressions? Even though the kind of embodiment which Merleau-Ponty outlines is a pre-reflective body, its lived and experiential nature means that embodiment cannot be known noumenally, as a thing in itself. It can only be known as a phenomenon which the human being experiences and lives through. While this is complicated by the fact that Merleau-Ponty defines phenomenology as 'the study of essences' – that is, the essential facticity 'of man [*sic*] and the world' (1962, p. vii) – what this amounts to is basically the sense in which the world's essence presents itself to us, as a fact for us, before we begin to reflect about it, thematize it and make sense of it.[9] On Merleau-Ponty's phenomenology, embodiment is known from the human perspective, but this is a perspective which is understood better by suspending it and by re-achieving direct and primitive contact with the world, via direct description of our pre-reflective actions and experience.

As we saw above, this leads Merleau-Ponty to celebrate the role of sensory experience and bodily expression in the formation of knowledge, but to ignore what our embodied experiences and expressions tell us. Shusterman's (2005) pragmatist take on embodiment agrees with Merleau-Ponty that our bodily feelings cannot be made completely transparent by our efforts to know them. However, instead of taking this as a reason to ignore them altogether, he argues that we should seek to reflect about our conscious somatic sensations because they are an important part of our lived embodiment which enable us to improve our bodily actions and understand them better.

Although these epistemological challenges remain largely unexplored by the other contributions we have discussed in this chapter, they are actualized by them. Our brief discussion of Behar's work suggested that introspective and autoethnographic reflection about inner feelings may constitute a significant basis for understanding how social reality is lived and experienced. Chicana feminism makes similar assumptions. In contrast,

Wacquant warns (too strongly in our view) against turning reflective accounts into a focus in social science research, but as we saw at the beginning of this chapter, his own research is not devoid of such accounts.

Embodied methodology

The inevitable question which follows from this concerns what techniques and practices we may draw on to investigate the bodily expressions, practices and felt experiences that make up the lived reality of embodiment. According to Merleau-Ponty (1962) the essence of human embodiment cannot be grasped through introspection or reflection without being distorted by those same procedures. Instead, lived embodiment should be investigated by observing and describing bodily actions, movements, gestures and appearances. Even intrinsic characteristics such as feelings and experiences, which Merleau-Ponty regarded as 'data of introspection', should be investigated in this manner because they too are expressed in appearance (p. 28).

Wacquant's (2015) call for ethnographic immersion and his scepticism towards excessively reflective projects reiterates some of these sentiments. As we will elaborate in Chapter 4, ethnographic fieldwork provides splendid opportunities to see what people look like and observe how they move about in a given space or place, how they express views and emotions through physical gestures, how they approach tasks through bodily actions and how they interact with others in face-to-face encounters. However, sustained ethnographic fieldwork will also give you a sense of what it feels like to live and work in a specific social setting, beyond what you see and hear. All this offers a rich basis for reflection as well as reporting: for reflecting about your observations and the felt experiences emerging from your involvement in the field.

Furthermore, immersing oneself in this way may open up opportunities for probing into research participants' reflections about their lived experiences – of work, of being part of a particular community, of being involved in a particular activity, of living under specific social and economic conditions, and of living with a particular body and identity. Ethnographic immersion may enable researchers to build close and trusting relationships with participants, where participants are comfortable with sharing such experiences and reflections.

This is not to say that other methods lack these capacities. Charlesworth's (1999) study of young people faced with poor economic and social prospects suggests that reflective accounts can be generated through interviews and conversations. In a different vein, Chicana feminism implies that any account of what it feels like to live a particular life within a particular sphere of society must be self-reflective. We will say more about embodied reflection in later chapters, when discussing ethnographic fieldwork, interviewing and diaries, and considering the processes and practices of analysis and writing.

Let us end this chapter by emphasizing that embodied research can take many different shapes and forms. Social scientists eager to conduct embodied research have a number of methods at their disposal, and we are positive that embodied research methods are particularly suitable for generating rich and lively accounts of what it is like to live and work in a particular social setting, under particular social conditions, with a particular body.

However, embodied research methods also pose a number of challenges. As we set out to elaborate these methods in subsequent chapters, we shall discuss their aims and practices as well as the practical and ethical difficulties they pose. Good embodied scholarship draws readers into the lives and worlds of others, often by revealing sensitive details about their bodies, their daily habits and their felt experiences. Hence, we are obliged to approach them with sensitivity and respect.

Questions and exercises

1. Why does Merleau-Ponty argue that inner embodied feelings cannot be investigated through introspection and reflection? What methods does he propose instead, and why?
2. Despite these challenges, what can be learned from paying attention to people's bodily reflections?
3. In what ways can embodied research offer insights into people's behaviour, interactions and social worlds?
4. Based on your reading of this chapter, what are the key components of embodied research? In other words, what can be learned from engaging in embodied research?

3

INITIATING AN EMBODIED RESEARCH PROCESS

WHEN SOMETHING STARTS TO ITCH

Quite often, it is our experience that new research projects start with an itch. That there is something which bothers us so much that we need to scratch under the surface to find out more. Or, that a more profound feeling or sensation triggers us to explore what that feeling is, what causes it, how it affects us and how it relates to other people. Whereas the vast majority of methods texts tend to offer rather rationalistic scenarios, recommendations or recipes for how to plan, prepare for and execute research projects, we think it is disingenuous to act as if we are able to maintain an objective distance and impersonal disinterest to the people and phenomena we study, or to pretend that we have a purely cognitive or intellectual interest in the research we are involved in. Embodied research necessarily involves passionate engagement, whether this is fired by sad or joyful passions.

When Torkild started his study on transgender in 2009 he found it impossible, and pointless, to try and detach himself and his own transvestite experience from the project:

> Torkild: At that point in time I had been out as a transvestite to friends, family and colleagues for more than 10 years, but I had never cross-dressed at work. For me, my transvestism was largely a matter of personal wellbeing and enjoyment, but I wasn't sufficiently comfortable about it to express it physically, on campus or at academic conferences. For quite a while I had been wondering what this was like for other transvestites. Were their experiences similar to mine, or completely different? Did they too struggle to express their transvestism at work? Did they struggle less than I did, or more, and why was it easier for some to be out at work?
>
> While I had met with other transvestites through a local festival, it had never felt appropriate to ask these kinds of questions, or to ask how they coped with their gender

identity at work and in daily life. So, when I decided to initiate this project together with my partner Louise Wallenberg, neither of us knew how to answer these questions. What we knew from the media, the internet and from previous research was quite limited. Some research on transsexual employees had started to emerge at the time; these studies (and certainly one study by Schilt and Connell from 2007) tended to emphasize how transsexuals adjusted to pressure from colleagues to do gender 'appropriately' and pass as a member of their desired gender category. But for me, transvestism had never primarily been about passing as a woman.

David has a somewhat different story to tell. In his case, much of his own gender trouble emerged when he discovered himself to be infertile, and this has at least partially been a driving force behind his research around gender and masculinity as well as his founding and developing of the journal *Gender, Work & Organization*:

David: I cannot say that the process had some kind of linear causal progression since much of the activity was a result of quite random serendipitous events. For example, when running the first Labour Process Conference in 1983 I was met by a barrage of angry feminists complaining that of the platform speakers, only one was a woman. I had made no special effort to recruit women speakers and in hindsight, this was naïve because in a gendered world such as academia, the men will dominate unless there is some intervention. The experience was not too pleasant, especially since as a fairly early career academic, this was the first conference that I had sought to organize. Indeed, it is a wonder it did not deter me from attempting the exercise ever again or even lead me to abandon academia for the more comfortable life I had experienced in the business world. Yet, after the conference and totally unrelated to it but probably due to my having recently completed a research project on equal opportunity for racial and ethnic minorities, I had an invitation to apply to the UK Equal Opportunity Commission to pursue research on sex discrimination in recruitment. This was successful and fortunately at the time I was supervising a research master's student who had a perfect fit to take up the research assistant post for this project. Gender then became a theme of much of my research largely because I was curious as to just how much the embodied experience as well as the dominant discourses of masculinity in our society had an impact on people's lives. At this time, though, I knew nothing of my infertility that would render masculinity a personal embodied trouble as well as a matter of public concern in our heavily heterosexually gendered and familistic world.

Let us not get too bogged down with these specific experiences. Our point is that much of our research has been significantly prompted and enabled by our personal embodied experiences. But it was also triggered by an interest in how this experience connected to other people, and how it related to existing research. Hence, the notion of embodied research that we are starting to sketch out here is not simply a matter of letting new projects be dictated by personal troubles and egotistic interests. After all, it is difficult to see how such troubles and interests may constructively inspire new research unless they are connected to something beyond that limited realm of our selves. As social scientists we are seeking to understand other people. But how well can we understand others if we have little grasp of who we are? As we try to connect with others, to find resonance or discord between our experience and theirs, do we not need to have some sense of our own experience too, and our own privileges?

At least since the feminist activist Carol Hanisch (1970) proclaimed that 'the personal is political', this has been a prevailing concern in much feminist research. But it also picks up on an underlying sentiment in the now classic sociology of C. Wright Mills. In *The Sociological Imagination* Mills (1959) argued that good sociology both engaged with 'personal troubles' and with 'historical problems'. According to Mills, the personal troubles that people live through in our relatively short everyday lives are not simply personal troubles, but deeply connected to and caused by problems and phenomena that operate at a larger scale throughout history. The economic hardship and poor health suffered by an individual or a family may be the result of large-scale technological and economic change which brings about massive unemployment. Equally, the anxieties suffered by a trans person or an infertile person may be related to social norms which privilege binary gender roles and stereotypes and nuclear family structures with a requisite number of biologically produced children.

Although personal troubles cannot be seen in isolation, they can also not be ignored. Indeed, Mills argued that the worst kind of social science at his time was exemplified by the 'grand theory' of Talcott Parsons and the 'abstracted empiricism' of Paul Lazarsfeld: whereas Parsons favoured abstract concepts and Lazarsfeld pursued quantitative causalities, they both did so to such an extent that the social realities they depicted were empty of social life and real people. To understand social life, Mills insisted, we need both a sense of biography and an understanding of history, and a grasp of how they are related: 'Neither the life of an individual nor the history of a society can be understood without understanding both' (p. 3).

As we shall see in later chapters, this threefold concern shapes the entire process of doing embodied research, including the early stages of initiating a new research project. But this also affects how we would like to speak about the initial phases of the research process. Research funding agencies, doctoral programmes and conventional methods texts very often speak of 'research design' – that the researcher planning a new project needs to develop and have a 'design' or 'structure' which depicts not only what the project is about but also how it is to be conducted and executed.

Sure, it will be difficult to undertake a research project without articulating its content and how you intend to carry out the research. However, the notion of 'research design' risks implying that a ready-made form can be imposed upon and implemented through a subject matter. As many scholars will admit to when pressed about the issue, research is hardly ever a linear process, and one rarely ends up where one started out. Our experience is that overly strong ideas and expectations about end results limit the potential to innovate and develop new and significant research contributions.

In spite of these qualms, we cannot ignore that the initial phases of embodied research involve doing certain things that most researchers do, regardless of topic, approach or discipline. It even involves doing those things in an approximate order. Admittedly, it is helpful to start off by carving out a research topic and a research problem, and to articulate one or several preliminary research questions early on in the process. It is also useful to think about what empirical material is needed to investigate those questions before deciding what techniques and procedures to use for generating and analysing the material.

Furthermore, there are questions about location and access that need to be considered at a relatively early point: Where will I be able to find the kind of material that I need to investigate my research questions empirically? How do I negotiate access to that location? What do I do if I cannot get the kind of access I thought I needed? Let us proceed by discussing these issues in a somewhat linear manner, whilst bearing in mind that, in practice, most of us tend to go back and forth between these stages as we struggle to develop a sense of what it might involve initiating and conducting embodied research.

EMBODIED RESEARCH TOPICS, RESEARCH PROBLEMS AND RESEARCH QUESTIONS

Insofar as a new research project is prompted or sparked by an itch or by a bodily sensation of something that troubles us, it is hard to imagine how a research topic or a research problem can be identified through a purely intellectual or cognitive process. But does that necessarily mean that embodied research needs to be explicitly about the body? Does embodied research require us to be concerned with the body as an object of knowledge and a topic of research as well as an active subject in the creation of scholarly knowledge?

As we saw in Chapter 1, there is no guarantee that research which is about the body will approach it in an embodied manner, even though we would prefer that it did. At the same time, it becomes difficult, if not impossible, to exclude the body as a topic in our research once we admit that it affects how we do research. Wacquant's (2004) embodied study of boxing, which we discussed in Chapter 2, may serve as a case in point. While Wacquant joined the Chicago South Side boxing gym with the opportunistic purpose of gaining access to observe 'the everyday reality of the black American ghetto' (p. viii) and 'the social strategies of young men in the neighbourhood' (p. 9), it soon became clear, through his own embodied participation, that both his own body and the bodies of his fellow boxers were a basic part of that reality which could not be ignored.

This is somewhat different in our own research. For Torkild the body has nearly always been an underlying concern. While most apparent in his research on transgender, there is also a personal background to his current research on wellness in corporate performance cultures, which is part of a joint project with Carl Cederström. For David, his decision to study the anxieties and insecurities of academics in a joint project with Caroline Clark was very much entangled with his own embodied experiences of entering academia. Let us elaborate, first with an account from Torkild, then with an account from David.

How Torkild came to study corporate performance culture

In my research with Carl we ended up putting the following aim in our research proposal: to investigate how managers and employees seek to manage wellness in organizations which are characterized by a strong emphasis on work performance. I have revised it since, a number of times. But how did we arrive at this initial aim?

From our previous studies in the area we knew that much attention had already been paid to the micro-political and disciplinary aspects of workplace health promotion; that despite employers' good intentions of wanting to help employees lead healthier and happier lives, such efforts involved quite invasive measures which could have unintentional and undesirable effects on employees and organizations. Reading the large literature on organizational performance cultures, we had also learned that such work environments were associated with stress and other health problems.

Luckily, we were able to identify a gap between these two literatures. As few efforts had been made to connect them, little was known about the specific ways in which managers and employees coped with problems of health and wellness within the specific context of organizational performance cultures.

This led us to pose the following research questions: In what ways do pressures to enhance wellness and performance conflict with or support one another? What social, discursive and bodily practices do managers and employees mobilize in order to manage and cope with wellness and performance? How may people commit to, comply with or resist dominant expectations of wellness and performance? And what unintended consequences may result as organizational performance cultures seek to manage wellness?

Despite the bodily flavour of these questions, this description gives the impression that we arrived at them through a purely cognitive and intellectual process of reading, logical reasoning and problematization. But would we have asked these questions regardless of how work-related health problems were reported in the media, or without any sense of how friends and family, colleagues and acquaintances experienced and dealt with performance pressures at work?

Perhaps it's too easy to come up with shorthand explanations when looking back at this process. Nevertheless, news reporting at the time had made us acutely conscious about how problems of health and wellness were connected to concerns about employee and corporate performance. In the summer of 2013, several international news outlets had reported about a 21-year-old intern at Bank of America who died after working 72 hours non-stop. Meanwhile, a number of Swedish companies were striving to increase physical activity among employees, their assumption being that fit and healthy employees would be more productive, more efficient and less prone to physical illness and mental health problems.

At the same time, we were stressing out about increasing pressures to publish in the right journals, to raise our citation figures, to maximize student satisfaction, and to demonstrate our relevance to business and society by appearing in newspapers and other media outlets. Then there's this: Just a few years before we developed the research proposal my mother had undergone open-heart surgery at the age of 57 in order to prevent the occurrence of a likely heart attack; at least in part, the cardio specialist had attributed her constant chest pain to the stress of working in a job with a high work load but little autonomy.

So, would I have delved into this topic had I not had 'body issues' of my own? If I hadn't had a history of worrying about my own body shape and gender identity? If I hadn't started to feel chest pain at the age of 32? If my sleep hadn't been interrupted by frequent worrying about work? If I could run without knee pain and go cross-country skiing without pressuring myself so much that my heart would start beating irregularly? Looking back, I suspect that these experiences (which were largely hypochondriacal but not entirely so) made me worry about these issues so much that I felt an urge to problematize them and explore them in a sustained and embodied manner.

How David came to study anxiety and insecurity among academics

Both Caroline and I had spent more than 10 years in industry after leaving school. Though we had benefited from this, we had always felt behind our academic colleagues.

I came to academia more by accident than design because, although starting work at 16, when I was 20 a friend likened me to the main character of Albert Camus's *The Outsider*. Although I had read nothing since leaving school, this was the beginning of my launching into a self-constructed, voluntary reading programme just for pleasure. My work at the time allowed me free afternoons when friends were at work so eventually I enrolled in a correspondence course to take 'A' levels, not for any other reason than to give my otherwise eclectic reading some organizational purpose. Then by accident I bumped into an acquaintance from school, Terry, who had just completed a degree course at university and after hearing about my studies suggested I apply since he believed they welcomed mature students. My exam results exceeded my expectations by a considerable distance, so I was readily accepted by the university to do a social studies course that was equally as eclectic as had been my early reading covering

(Continued)

(Continued)

anthropology, economics, geography, philosophy, politics, psychology and sociology. Because my wife was planning to give up work to attend college and due to having a large mortgage on favourable terms from my employer, I continued working full-time and since the degree was a hobby I enrolled clandestinely from my employer. The result was that two full-time activities put considerable pressure on my body and the resulting neglect created a number of health problems that began to plague me more and more as anxiety about meeting deadlines accumulated. In addition to routine duties, my formal work involved sales targets and pressures to perform that were stressful, especially as I was seeking to avoid my academic activities being disclosed. Eating regularly was sacrificed and I ended up with considerable stomach pains eventually being diagnosed with ulceration that was on the verge of being operated on if I failed to take greater care of my body. This I tried but the condition was not helped by my increasing anxiety about 'winging' my academic studies and comparing myself with a student colleague on the course who stayed with us and seemed always to be working. His anxiety must have been even greater than mine, however, because perhaps due to predictions of securing a first-class degree, he disappeared halfway through, although news of his suicide did not break until after the exams.

I went on to do postgraduate studies at a different university but in management, so I was thinking vaguely that this might either enhance my prospects at work or stimulate a change of career. By this time, my employer had heard about my extra-curricular activities and offered me a major promotion at their head office that I turned down in order to continue with my studies. While I feared that this risked getting me sacked, I had taken the precaution of setting up my own business that could readily replace the lost income if my employment was discontinued. This resulted in me doing three jobs, which was not great for my troublesome bodily health although later it did improve. In many senses, I was still treating university work as a hobby partly because I never felt 'good' enough to be an academic. When, due to supervisory advice, I began to favour a switch from business to academia I still attributed securing my first lecturing post as less due to my academic competence than to the head of department being impressed by my business background.

Though it is only in recent years that I have started to investigate the insecurities and anxieties of other academics, these are issues I have struggled with throughout my career. Interestingly, I discovered that the sense of being an imposter was much more widespread, for in the study with which I opened this little autobiography, we found it to be one of three very common feelings of anxiety and insecurity among academics in a culture of increasing surveillance, performative demands and work intensification (Knights and Clarke, 2014, 2017; Clarke and Knights, 2015).

GAINING EMBODIED ACCESS

Once you have established a research problem and a set of research questions, most research guides and methods texts would recommend that you start thinking about what empirical material you need to answer these questions, where you might find it, and what techniques and procedures will enable you to generate it. Moreover, they are likely to recommend that you have a sense of what time frames and resources you have available before you start contacting people and negotiating access to specific locations. If a project is planned as part of a course or a thesis, there will be time limits but also time available. If you have an academic teaching job, you may need to generate funding which enables you to reduce your teaching load.

What we are suggesting here is not that different, except that the kind of methods we would tend to endorse involve an embodied engagement and presence which is rarely made explicit in conventional methods texts. We will discuss a range of specific methods in Chapters 4 to 8. But before doing so we need to say something about the challenge of gaining access to sites, organizations and communities which allow you to be so present with your body that you may generate empirical insights into the embodied lives of the people inhabiting these settings. Again, we are not suggesting that there are any simple ways of resolving these challenges. Rather than presenting a list of recommendations or a number of success stories, we will share some of our own failures and difficulties. Let us begin our exposé of this long journey by sharing Torkild's experience when he and his colleague Carl started to formulate their project on wellness in organizational performance cultures.

How Torkild worked his way into a corporate performance culture

Carl and I had no research grants and our teaching loads were too heavy to allow us to do what we were planning in whatever time we had left. Our only option was to apply for research funding. Like most academics, we worked long hours, late evenings and weekends to develop our research proposal. We pulled our hair in frustration over strict formats and tight word limits, rubbed our sore eyes which were dried out after hours in front of the computer screen, and worried that our aching backs, necks and shoulders would never recover from the process. Months later, we opened decision letters only to find that our application had been rejected funding.

What did we do? After a depressing few days dominated by a sad combination of moaning, alcohol consumption and rage against the idiots who didn't understand our genius, and a longer period when the last thing we wanted to do was think about the project, we were eventually back at it, revising, rethinking and rewriting the application. We got rejected again, at least twice. But we got lucky the third or fourth time. Excited about the money and the freedom it gave us – to think, to research, to write – it soon dawned upon me that I needed to get on with my part of the project.

(Continued)

(Continued)

Carl had already started doing interviews for his part, which focused on the use of performance-enhancement drugs in working life. I, on the other hand, thought I'd been promised terrific access to do ethnographic fieldwork in a Big Four accounting firm: to hang around in their offices, to observe people as they went about their daily work, to talk to them and have lunch with them, to sit in on meetings, recruitment activities and staff development workshops, and to participate in various fitness activities. This seemed ideal because Big Four accounting firms are widely regarded to have strong performance cultures: productivity targets are high, staff are ambitious and they know how to measure things; they attract people who are hungry for success, and they promote the few who come out on top.

A couple of years earlier I'd done a pilot study of the firm's efforts to improve employee health. Along with my research assistant, Carl Lundevi, I had done several interviews with employees, and I had gone running with them on a regular basis. In fact, my main contacts at the firm had suggested to me that getting their employees to come running every Monday at 4.30pm was a terrific way of tempering the performance culture. Their idea was that taking a break from work to go running with colleagues was good for business. Even though the running would take time away from work and reduce short-term productivity, they were optimistic that it would enhance productivity in the long term because it would prevent employees from getting stressed out and burned out.

I got in touch with my main contacts a few weeks after we got the funding, keen to take them up on their previous offer to expand the research project. We met for lunch. My main contact was positive and sounded excited about it all. He had already talked to his Vice President, who apparently had liked the idea, but asked him to 'check with HR'. My contact told me to draft a letter which he could present to the HR Director. He quickly got back to me and asked me to make a couple of minor revisions. I did as he suggested and sent off the new version.

I started waiting. One week, two weeks, then four weeks. I heard nothing from them. I contacted them again, suggesting modifications to the project which would make it less of an encroachment on their time. My contact promised to get on to HR. I waited another two weeks, and then the email arrived:

> Although we are excited about the opportunity to do this project with you, and our HR Director sympathizes with the research approach, she says that there are too pressing changes going on at the moment which need our full attention. ... An opportunity to do the project might however arise sometime in the future. ... Best wishes

Part of me wanted to cling onto that bit in the email which said 'sometime in the future'. But I somehow knew that was a polite euphemism for 'never'. I was in despair. Would I have to give back my part of the grant to the funding agency? Would I be able to use the data that had already been generated about the firm to explore the issues I'd promised to do in the research proposal? Should I change my topic and hope that I would still be able to say something about wellness and un-wellness in performance cultures?

My collaborator Carl Cederström had a suggestion: 'Torkild, why don't you get in touch with that sportswear company where they've made physical exercise compulsory? … I think the CEO may be keen.'

I was hesitant. I thought they'd be too extreme, that they believed too strongly in a causal link between fitness and performance, and that the nuances I wanted to explore would just not be there. I didn't want to reinforce simplistic stereotypes and believed that a stronger contribution to research could be made through a case where managers and employees were more ambivalent about the connections between wellness and performance. How I could let my own prejudices affect me in this way, I would rather not dwell on, but it does feel stupid in hindsight.

Luckily, my desperation reached a level where I had little other choice but to go along with Carl's suggestion. Five months after the funding agency had started making monthly payments into our project account Carl sent off an email to the CEO about the possibility of interviewing him for the project. We got an almost instant reply from his management assistant, who told us that the CEO was keen to meet us after the summer holidays. 'Wow, this might work!' we told ourselves enthusiastically. We agreed on a preliminary date, and to get back in touch in late August.

The days leading up the meeting I was quite nervous. A lot was at stake: the project, but also this book. The book would certainly benefit from more first-hand experience. So, this time it just had to work out. But what exacerbated my nervousness was the CEO himself. Before the first meeting I had read up on him online. He was my age but had so much which I didn't. The media described him as a fast thinker and a fast talker. He had a hugely impressive corporate track record, a lean and muscular body, and the self-confidence to go with it.

The time for our appointment came. We showed up at their head office 10–15 minutes early. My heart beat faster than normal. I was sweating. My palms were wet. We rang the doorbell and were let in through the glass doors by a friendly receptionist. We introduced ourselves and told her we had a scheduled meeting with the CEO. 'Welcome! Would you like a drink? Coffee? Water?' Carl declined her offer. My mouth was drying out, and I somehow managed to say yes to a glass of sparkling water, without seeming too desperate. We sat down on a sofa, looking at each other in silence.

After five minutes the CEO came down the stairs. He was wearing smart trousers, no socks in his swanky leather shoes, and a slim-fit shirt, buttoned down and perfectly ironed. Just like I had seen him in some of the photographs:

'Hello, lads! How about a game of ping pong?'

The ping-pong table was standing next to us. I didn't know where to hide. I had barely played at all, in any case it was ages ago, and I was useless at sports, particularly racket sports. Carl volunteered, which was just as well. Before I knew it, Carl had won the game. The CEO blamed the bat and walked us into the fish tank next door while I was doing my best not to drop my rucksack, my cycling helmet, my mobile phone, my notepad, my pen and my glass, which was half full of sparkling water. The CEO sat us down at a small round table and asked us to tell him what we had in mind.

(Continued)

(Continued)

Carl did the talking, making sure to mention that we had received generous funding from the research council, and the fact that he had recently published a piece in *Harvard Business Review*. The CEO seemed impressed. Carl emailed him a link to the paper. He looked it up on his mobile straight away, nodded and smiled.

'So', Carl said, 'we were hoping to do two things: firstly, we're planning a series of articles on this topic for *Harvard Business Review*, where we meet with a number of Swedish CEOs, individually ... we first join you for a work-out, and then we take you to luncheon, for an interview, you know, kind of like that thing in the *Financial Times*. ... But ... Torkild in particular would also like to do an ethnographic study, if possible, of how companies with strong performance cultures manage wellness.'

The CEO now looked at me. I stuttered and tried to say something about my previous research on the accountants. He seemed less impressed with this and started talking about what his company does to improve employee fitness and performance. Nevertheless, I picked up the words, 'that sounds interesting ... I think we would like to do that ... Let's do lunch next Friday. Then you can come to our sports hour at 11 ... at the gym, just around the corner from here ... But our HR Manager needs to okay that other study ... I'll have my assistant set up the meeting, with HR too. Now, I'd like another game. But let me go and get my own bat upstairs.'

The CEO won the second game. I managed to take a few photographs with my mobile. We shook hands, said thanks and goodbye, and left the premises. Once outside, Carl and I looked at each other. I had difficulties containing my excitement. As soon as we had turned around the block, I showed Carl the photos. We picked one, which I immediately emailed to the CEO. Instant reply:

'Thanks! The bat made a difference. See you soon.'

The following Friday we showed up at the gym as agreed. Dozens of employees were streaming into the room. The CEO barely said hello to us but told us to meet him at the Asian restaurant next to their head office straight after the work-out session. Warming up, the coach had us all jogging sideways in a circle around the room. This was an awkward activity for me. I felt a sudden pain in my knee and immediately realized I had strained a muscle. What did I do? I carried on of course! My project depended on it.

After an hour of what was my hardest work-out in years, possibly ever, we hurried to the changing rooms, took a quick shower and got changed as fast as we could. The CEO was already out of there while we were dressing. We managed to find the restaurant, and the head waiter took us to the CEO's table. We ordered the same as him, even the same kind of diet coke, and started the interview. Carl asked most of the questions, and the CEO talked on and on: about his own background, his leadership style and the company. He had initially agreed to give us an hour. After an hour and twenty minutes he asked what time it was:

'Nearly 1.30', I said.

'We better get going then! My HR Manager will be waiting for us.'

Back at their head office the CEO introduced us to the HR Manager and disappeared upstairs. She walked us into one of the showrooms and asked us to tell her about the project.

Again, Carl did most of the talking, but stressed that I would be doing most of the actual research. I felt more relaxed now and told her that we were interested in finding out what it's like to live and work in a performance culture, particularly one where management seeks to integrate employee performance and wellness. The HR Manager responded with enthusiasm, mentioned that 'it may be good to get an outside perspective on what we're doing', and offered us to participate in a number of different activities:

'You'll be free to hang around in the reception area whenever you like … we do weekly meetings every Monday at 9pm. Make sure to be there on time because the CEO hates tardiness … then you should come to our weekly sports hour, of course … then we do staff development workshops that I can invite you to … we try to give our employees the tools so that they can achieve focus in their daily work … we do introduction sessions for new staff … but we also do workshops on stress management, on feedback …'

I could hardly believe what I was hearing. The access to all these activities.

'Just send us a proposal about the project', she told us. 'Email it. To me and to the CEO. Then we'll present the project to our employees on Monday, and if they're fine with having you around then you can come along the following Monday.'

That afternoon Carl convinced me to celebrate with a pint at the pub just down the road, though I admit I would have preferred somewhere further away from head office. The HR Manager had mentioned that they didn't see any value in after-work beers, and I didn't want her or anyone else from the company to spot us sitting outside that pub drinking.

The following morning, I drafted the letter and passed it on to Carl. My knee was worse than the previous day, but I hoped the pain would go away after a few days. We passed the letter between us a couple of times over the weekend, and when it was down to a couple of pages I sent it off. We had made sure not to ask for too much. Knowing that the company was eager to optimize performance, we emphasized that we wanted to interrupt their ordinary work as little as possible. Instead of focusing on formal interviews we asked for permission to sit in on workshops, meetings and other activities, to hang around in the reception and the lunch room, and to engage in informal conversation with staff during lunch time and coffee breaks.

At the same time, management's enthusiasm had made us optimistic that if we played our cards right, we might be allowed to do more things further down. We made sure to promise that we would follow the ethical guidelines of the Swedish Research Council, that we wouldn't reveal individuals' names and identities, and that we'd offer regular feedback on our preliminary findings to management.

The following Tuesday morning the HR Manager had already given us the okay via email, and a series of invites to various activities started streaming into the inbox. Not even in my wildest dreams had I fantasized that a company would provide me with this kind of access. Hence, I'd already booked up a number of Fridays with other things, work trips and leisure trips. Thankfully, I was able to accept a number of the invites, and made sure to send her an email where I extended our thanks for their generosity and apologized that I wouldn't be able to participate in all of the most immediate activities she had invited us to. In secret, I worried that my knee would be in such a bad state that it would take me weeks before I could come back to their sports hour.

We shall draw more on this research project in subsequent chapters, particularly in the next chapter where we discuss how embodied research can be conducted through participant observation, shadowing and other ethnographic methods. After this exposé, let us now try and summarize the key lessons we take away from our mixed experiences of initiating embodied research projects.

SUMMARY: FROM EMBODIED PROBLEMS TO EMBODIED ACCESS

By contrast with conventional methods books that tend to depict the research process in largely cognitive terms, in this chapter we have drawn attention to the embodied aspects of formulating research problems as well as the challenges involved when seeking to gain access to do embodied research. Ordinarily, a new study acknowledges prior research in the field, but it also requires some personal engagement. To paraphrase Mills's (1959) notion of the sociological imagination, the formulation of new research problems may be prompted by an embodied experience of personal troubles but it still needs an intellectual sensitivity to public issues and an understanding of pressing social problems.

This may take social scientists into a variety of projects, questions and empirical settings, and in this chapter we have sought to illustrate the challenges and difficulties of initiating new research projects by drawing on some of our own experiences. Doing so, we have focused on the stressful experience of developing research proposals and grant applications, the distress and disappointment associated with the failure to gain funding and empirical access, face-to-face meetings with contacts and possible research participants who may or may not lead to access, and bodily involvement in the stages of negotiating empirical access. Indeed, securing quality access is a wholly embodied process since it is through the embodied engagement of the researcher that gatekeepers will warm to and help facilitate the research. Our discussion has even actualized other questions, such as what access to ask for and how to make a project seem interesting or beneficial to research participants.

However, since research is hardly a linear process, these questions are rarely resolved before the researcher has been admitted to a research site. Rather, they tend to be continuously negotiated throughout the period when you are generating empirical material. This is particularly the case if you are seeking to study other people's bodies, bodily interaction and sense of embodiment, which is such an intimate part of who we are.

Although ethical guidelines regarding informed consent, participant anonymity and feedback offers basic ideas of how to address the bodily integrity of research participants, it is difficult to know up front how to adequately acknowledge the sensitivity, vulnerability and power of the people we involve in our research. What we have learned through our careers is that there will always be tensions, contradictions and ethical challenges which require our constant engagement. We will return to some of these issues in the next chapter, when discussing the challenges of remaining embodied while doing ethnographic research.

Questions and exercises

1. Think about an ongoing or previous research project or dissertation project that you have been involved in:
   a. In what sense was your research problem triggered by the embodied experience of a personal trouble? If you struggle to remember or identify such an experience, think about why you struggle to do so and how you came to arrive at your research problem.
   b. Assuming that your project required you to negotiate access to an empirical site, in what ways did you seek to manage your body and bodily demeanour while doing so?
2. Based on your reading of this chapter, mention the main challenges of initiating embodied research.
3. All social science research involves ethical challenges and dilemmas. What particular ethical issues may need to be considered when initiating embodied research?

4
EMBODIED IMMERSION AND ETHNOGRAPHIC FIELDWORK

PHYSICAL PRESENCE, EMBODIED KNOWING

The embodied aspects of research are particularly apparent in ethnographic fieldwork. Even in contemporary society, where much human activity is mediated by information technology, researchers doing fieldwork need to be physically present – to observe what other people are doing, hear what they are saying, and experience what it is like to live and work in their surroundings. How could we otherwise participate in their everyday activities, become part of their culture, and share their worlds?

No doubt, there is growing consciousness amongst ethnographers about the bodily practices that are involved in fieldwork and the embodied nature of the ethnographic research process. As Yanow (2012) has pointed out, the researcher's body is 'the main instrument of ethnographic knowing' (p. 33), and there is an emerging literature on how we use our bodily senses when conducting ethnographic research (Stoller, 1997; Pink, 2015). Furthermore, leading commentators have argued that social scientists doing ethnography need to become bodily immersed in the settings we study (see especially Wacquant, 2015).

In this chapter we shall discuss some of the main opportunities and challenges that ethnographic research methods pose to the embodied research approach we pursue in this book. Doing so, we shall engage with distinct voices in the ethnography literature, but also revisit Torkild's study of corporate performance culture in the Swedish sportswear business and share some of David's experiences from doing fieldwork in the UK newspaper industry.

THICK DESCRIPTION AND AUTOETHNOGRAPHIC REFLECTION

While this chapter will primarily engage with ethnographic fieldwork as a means of generating rich empirical material, we are not trying to reduce ethnography to a method of data generation. Neither of us can claim to be a professional ethnographer, but we do appreciate what its etymology tells us: that ethnography means 'writing culture', and that ethnography is an entire scholarly enterprise wherein fieldwork is closely entangled with the process of writing. Thus, the written accounts we present from our own fieldwork are not given or simply recollected, but sampled and selected, construed and reconstructed through an embodied engagement with ideas, interpretations, feelings and emotions.

As Clifford Geertz (1973) pointed out at the beginning of *The Interpretation of Cultures*, ethnography is not defined by its techniques and procedures. Although ethnographic fieldwork involves 'establishing rapport, selecting informants, transcribing texts, taking genealogies, mapping fields, keeping a diary, and so on ... it is not these things, techniques and received procedures, that define the enterprise' (p. 6). 'What defines it', Geertz continues, is its 'thick description' (p. 6); that is, all the interpretation which is required to create written accounts of culture, to decide what to include and exclude, what cultural patterns to emphasize, and how to construe a context which makes the account meaningful to its readers.

It would be disingenuous to silence the interpretive work of thick description or abstain from the reflexive challenge it poses. Since our fieldwork accounts are as constructed as our post-hoc rationalizations of them (Clifford and Marcus, 1986; Wacquant, 1995), we are urged to reflect about how we are involved in the cultures we write about, and how we decide to write about them. We shall say more about writing in Chapter 8, but it will become apparent from the accounts we share in this chapter that they are constructed to flesh out and nuance the embodied practices, experiences and challenges of doing immersive fieldwork within a particular setting, whether a contemporary performance culture or the UK newspaper industry in the 1970s.

Thus, there is an autoethnographic dimension to all ethnographic research, which turns us into objects as well as subjects of enquiry, into observers of our own observing (Ellis, 2016). Despite the 'auto', autoethnography is distinct from autobiography: its object is cultural experience rather than personal experience (Holman Jones et al., 2016). The autoethnographic dimension of ethnography, then, challenges us to reflect about the cultures we have become part of and what it feels like to embody them; to reflect about how we embody those cultures rather than our personally embodied experience as such.

There is an ethical dimension to this too. As we examine, reflect about and describe our experience of being embodied and embedded in a culture, our own lives are open for interrogation by ourselves and by those who might read us. While our readers might judge that we have embodied a culture poorly or lived a bad life within it, we may endeavour to change our lives by thinking critically about who we are, 'how we want to be' (Ellis, 2016, p. 10) and how we embody the cultures we immerse ourselves in.

OBSERVATION, PARTICIPATION AND IMMERSION

Recognizing these ethical and epistemological dimensions does nothing to downgrade the concrete practices of fieldwork. The fieldwork methods of participation and observation have been fundamental components of ethnographic research for a long time, at least since Bronislaw Malinowski started to cultivate participant observation as the primary method of anthropology in the 1910s and 1920s. Already during his fieldwork among the Papuan inhabitants of the Western Pacific Island of Mailu, Malinowski developed his influential view that ethnographers should live with the people they studied over a long period of time in order 'to grasp the native's point of view' (Malinowski, 1961 [1922], p. 25).

While the difficulty of grasping the native's point of view may be best testified by Malinowski's (1988 [1967]) own diary, which was published posthumously and included a number of rude comments about the people he studied (see also Geertz, 1974; Wacquant, 1995), it does not undermine the value of fieldwork. Certain commentators regard field immersion as the *sine qua non* of ethnography, as is illustrated in a rather uncompromising manner by the organizational ethnographer John Van Maanen (2011): 'If one cannot do lengthy and sustained fieldwork among others who are often initially recalcitrant and suspicious of those who come uninvited into their lives, one has no business doing ethnography' (p. 219).

Meanwhile, the relationship between participation and observation remains an unsettled issue. Different kinds of ethnographic research involve different degrees of participation and different kinds of participation: from relatively passive forms where the ethnographer is largely a bystander or 'fly on the wall', to more active forms where the ethnographer is deeply involved and bodily engaged in the same activities as the people they study (see, e.g., Spradley, 1980) and even live with them for longer or shorter periods of time (e.g. Malinowski, 1961 [1922]; Goffman, 2014; de Rond, 2017). Either way, presence implies some kind of participation, and even bystanders are likely to affect the situations they observe.

In practice, the degree of participation and observation is likely to vary immensely during the course of fieldwork, difficult, as it is, to be equally devoted to observation and participation at one and the same time (see, e.g., Gleason, 1989; Aull Davies, 2008). It may even fluctuate during one and the same field trip. Moreover, how much you participate in various activities is likely shaped by what kind of empirical access you have been able to negotiate. This has certainly been our experience. As Torkild recollects:

> In my ethnographic fieldwork at the sportswear company I found myself very much in a bystander role during weekly morning meetings, but I participated like ordinary employees in the company's weekly sports hour. To other meetings I'd been invited to attend as an observer, but sometimes found my role changing without notice, as staff members and managers asked me, on the spot, to offer comments and feedback on what they were doing.

In David's case, this has been rather different, because he has primarily pursued an immersive approach in his research on academics in UK business schools:

> In studying academics after having been one for over 45 years, I was completely immersed in the practice. However, I also had to stand a little apart from it in order to conduct interviews. The publications that came out of the research explicitly drew more from the interviews than from the embodied experience of the work, because we needed some external and more public evidence of what we had internally and personally experienced over a number of years. At the same time, it is unlikely that we would have achieved the kind of reflexive 'openness' that we desired concerning the lives of our participants had we not ourselves suffered the same anxieties and insecurities of being academics.

Insofar as we do ethnographic work, then, we constantly find ourselves in situations where we have to negotiate our dual roles as outsiders and insiders, whether we study other people's cultures (see, e.g., Spradley, 1980; Woodward, 2008) or our own cultures. In our experience, these challenges are especially acute in research processes where we are eager to study other people's bodies while seeking to develop an awareness of our own embodied involvement. But there are even concerns related to bodily presence in ethnographic fieldwork that lacks an explicitly embodied research agenda. Let us address these issues further, first in relation to Torkild's fieldwork in the Swedish sportswear business, and then through David's study of a sales department in the UK newspaper industry.

Embodied immersion in a sportswear company

As we mentioned in the previous chapter, Torkild had been given quite generous access by top management to study the company's performance culture. Despite all the possibilities this opened up, it even raised a number of qualms:

> While I had promised to primarily act as a 'fly on the wall' in order to disrupt work as little as possible, I had also been invited to participate in a number of activities and to engage in informal conversations with employees during lunch breaks and coffee breaks. This made me optimistic that I'd be able to follow Wacquant's (2015) appeal and become deeply immersed in the organizational culture.
>
> However, I also recognized that I had to stay sensitive to their needs if I were to maintain, and preferably increase, my empirical access to the organization and to the people working there. In particular, I didn't want to come across as a spy or as a 'peeping Tom' who encroached on their privacy, and I didn't want to add stress to what I suspected was an already intense work environment. But, I also wanted to get to know them as well as possible. How would they otherwise be prepared to share their everyday experiences, and tell me what it's like to live and work in this culture?

To see how these qualms and concerns played out, let us go back to the time around 16 September 2016 when Torkild attended his first team meeting.

Torkild's first day at head office

In the first meeting Carl and I had with the company, management had stressed how important the weekly sports hour is, not just for improving employee fitness, but for team building and for strengthening the performance culture. But the HR Manager had also mentioned their weekly 'team meetings', which are held every Monday morning at 9am sharp.

Come Monday, 16 September. Carl and I make sure to be there at 8.45am. Striving for a clean-casual balance, I'm wearing a pair of khakis and a striped cotton shirt, only to find that Carl is wearing something very similar: 'We have to stop dressing like twins', Carl utters as we enter through the front gate, crushing my hope of fitting in. Once we're buzzed in through the glass entrance we follow a hoard of mostly young-looking and casually dressed employees streaming into the kitchen. At the same time, I struggle to think of good questions to ask and appropriate ways to initiate conversation or break the ice. The HR Manager says hello and we manage to introduce ourselves to a couple of employees. We've agreed earlier to sit at two different tables in order to try and maximize our reach. The kitchen area is filling up fast. By 8.55am there are probably 50 people in the packed room, possibly more. I'm no less anxious this time, but I find a seat next to the Logistics Manager. I had been paired up with him at my first sports hour, and we had got along well. I even manage to introduce myself to a couple of other people at the table, including a guy who turns out to share my interest in cross-country skiing.

All of a sudden, a loud voice cuts through the background noise of chatter:

'OK, guys, it's 9 o'clock! Let's start!'

The CEO has walked up to the front of the room. His fitted shirt, slacks and leather shoes give him a smart-casual but expensive look. His presence by the flip-over is formidable. He quickly picks out one woman, who started working for the company two months earlier. He asks her to come over to present their framework.

'So, what's the first question we ask ourselves?'

'Where are we?', the young woman hesitates.

'No! It's where are we going, 'cos if we don't know where we're going we don't know where we are. So, where are we going?!'

While he scribbles something on the large piece of paper she starts mumbling something about total sales, but is quickly interrupted again:

'Yes, sales are important, but our first KPI [key performance indicator] is sell-through. So, what's our sell-through target?'

'60 per cent?' she tries.

'No! Please help her out will you.' He has turned to the floor and gets a unison in response: '70 per cent!'

'That's right. If we're gonna reach our targets for 2019 we need to aim high you know.'

He hands her the felt pen and tells her to continue. With firm guidance from the CEO and fellow employees she is now able to put down all the KPIs as well as all the questions making up the company's performance framework. Together, they fill in the current performance figures too. Something is red, something is green and something is yellow. Some other thing is nearly green but actually red. The CEO thanks her as she walks back to her seat:

'See, you did this much better than me. You were more pedagogical. I talk too fast. And loud ... But I'm here to guide you and get you excited!'

The CEO continues the meeting, asking team managers about last week's performance and this week's tasks and events. Their voice and posture speak of a mostly self-confident group. Even though the CEO keeps interrupting pretty much everybody he asks to talk, the meeting proceeds swiftly. They all seem to know the drill: stay focused, present the facts and express excitement. I strive to jot down as much as possible of what is said. Since it's my first meeting everything is new to me: their framework, the jargon, the people. I have little idea who is who in the organization, and I struggle to hear what they say and figure out the significance of what they say.

Concluding that overall performance is better than last year but behind budget in nearly all key areas, the CEO gives off a serious but energized look:

'You know, I'm here to help you allocate your time and resources better!'

He grabs another marker, turns the page on the flip-over and starts writing. I'm sitting at the back of the room and find myself having to squint to be able to see what he scribbles down:

'Some things are important', he states, 'especially that we reach our targets by 2019. Other things are urgent and important, such as sorting out that logistics issue – otherwise our customers won't get our products and if they don't then we won't get paid.'

He moves on, and I finally realize that he has drawn up a two-by-two matrix to show employees how to prioritize between different tasks.

Suddenly, he asks people to offer feedback to each other. (I later find out that the management team had a feedback session the week before.) Most staff seem rather reluctant, except for one woman, who has already spoken, but he brushes her off, smiling, that he wants to hear from someone else for a change. Eventually, the CFO says something, addressing the Logistics Manager who sits next to me:

'I'm really impressed with how well you have started handling issues with our supplier. It's more based on facts now, so they listen, and you get it sorted.'

'How did that feel?', the CEO asks.

'It felt good', the Logistics Manager admits.

'We have to celebrate the successes you know! Anyway, let's get to work!'

It's 9.35am and I realize the meeting is over. People get up to leave, some heading fast out of the room, presumably back to their desks, while others get refills of coffee and water before leaving. Most of them ignore me. I find Carl across the room. 'This is interesting stuff isn't it!' I whisper before he reminds me that he'll be off and that I'll be doing the fieldwork on my own while he'll be concentrating on his own study.

(Continued)

(Continued)

A couple of employees stay on in the kitchen area, chatting quietly, while I move over to the reception area, not wanting to be seen as an eavesdropper. I have a quick look through my notes and try to add things that I missed to write down during the meeting. Eager to get as much material as possible, and anxious not to miss anything, I start writing down pretty much everything I see and hear: details about the room, employees chatting, staff and visitors coming and going, pop music playing in the background, the big screen playing the company's ads (this is the first thing people see as they enter the front door). For my entire academic career, I've felt that I've lacked rich empirical material. I've never done a proper ethnographic field study before, and I'm determined not to waste this opportunity.

The chair I've resorted to is soft and low and probably not great for my back. But it gives me a good spot to observe staff from, as they need to pass through the reception to get to the kitchen, the showroom or to any of the meeting rooms on the ground floor. Some staff ignore me completely, while others say hello. A couple of people even tell me that 'It's great to see you here!' One of the staff members I talked to earlier cheerfully asks me why I don't have a proper desk yet. Head of Marketing comes by and introduces himself. What a nice guy, I tell myself. Later on, a former student, who I don't recognize, comes by and says hello, but is quickly called on to hurry up by someone who seems more senior. They're having a meeting. Meanwhile, the receptionist is clearing the kitchen, taking calls, buzzing visitors in and tending the coffee machine. The Logistics Manager walks by smiling: 'From one meeting to the next.' The CEO ignored me during the meeting, but comes by on his way out:

'How's it going?'

'Very well thanks. Great to sit in on your meeting. High speed', I manage to say.

'We do this every Monday you know. Just come along.'

A casually dressed woman in her 30s comes in and parks her bike next to where I'm sitting. She says hello and I introduce myself. She had missed the morning meeting. Later that day during lunch I bump into her again, which leads to a surprisingly long conversation about my project, about the company, about the fitness activities and about the performance culture. She never comes to the sports hour anymore, she tells me, because of permanent sports injuries she developed when she was younger:

'I went to a couple of ones when I first started here, but they gave me so much pain that I ended up spending the entire weekend in bed, on pain killers.'

This employee later turns out to be an important contact: she is talkative, she is incredibly friendly and she seems to keep some distance from the corporate culture.

At 11.30 I realize it's lunchtime as three or four employees start heating up food in the microwave. I ask the receptionist about lunch places in the area. Knowing the company's focus on fitness I don't want to show up with a grilled sandwich dripping of fat, so I end up getting a too expensive salad with fresh bits of tuna from the gym around the corner. (Given my own blood sugar issues I know I need to fill up with proteins.)

Back at head office, and dreading rejection, I walk up to the only occupied table and carefully ask if I may join them: 'Of course', they reply politely but without enthusiasm. The three women, who all seem to be in their mid-30s, are in the middle of their homemade meatballs, pasta and salads.

'You're the early bunch', I try.

'I've been here since 7.40am. And you're doing that study with us, right?'

While I do my best not to gobble up my food like I usually do, they continue chatting as if I'm not there – about rush-hour traffic, about when to best drop off and pick up their children from kindergarten, about getting their life puzzle sorted. Even after the HR Manager and the CFO have joined – she with a homemade plate of tomato-sauce spaghetti and he with a big juicy burger from the courtyard food truck – work is hardly mentioned. Rather, the conversation turns to food and eating habits, and how crickets will soon become the new big source of protein.

People come and go from the table, but I stay, eating more slowly than usual, smiling along, trying to ask a follow-up question here and there whilst being careful not to interrupt the conversation. After an hour or so I'm the only one left when the HR Manager gets up to leave. She moves fast and avoids eye contact. Then again, this is supposed to be a performance culture, and she's probably got loads of work to do. I remind myself that I need to try and make contact with as many people as possible, and that I need to earn their trust. I can't come across as a management lackey.

I decide to get a cup of coffee from the machine and start looking across the room for a table to join. Next to the coffee machine is a long table with 10 people. Fearing rejection, I decide against it. Across the room, I spot the Logistics Manager I talked to earlier, but he seems busy conversing with four people I suspect are in his team. He certainly gives me no indication that I should join them. I finally decide to walk over to the next table, where one woman has just started her lunch.

'May I join you?'

'Yes, please. Who are you?'

As I give her a two-sentence spiel about my project, I notice that someone I talked with in the morning is sitting next to me. She puts down her mobile and joins the conversation. We talk about what they do and what they did before joining the company. Even though I've only just briefly mentioned my project, one of them gives me a rather detailed view on the company and what it's like to work there: on the sports hour ('It's not for me so I don't go to it'); on the CEO's masculine leadership style ('It bothered me immensely at first that he referred to all of us as "lads" but he has calmed down a bit'); on what it was like before the new management team took over ('It was fun and cosy, but we didn't make enough money and the last interim CEO had no clear mandate so he didn't know what he was doing'); and on the performance culture ('Now that we've got measurable individual targets every one of us actually knows what to do').

In the meantime, a couple more people have joined the table, and when they all leave to get back to their desks, it's 1.15pm. I go and get my rucksack from behind the reception, and I'm taken aback a bit by how friendly the receptionist is. She is almost treating me as if I work there. She tells me how to access their internet and I stay on for 45 minutes or so, finishing a reviewer's report that I'm two weeks late with. At 2pm I say goodbye and try to walk as swiftly as I can to catch a bus back to campus. To my relief, my knee doesn't actually hurt that much. As I enter the department I'm just two minutes late for my meeting with the teaching committee.

Reflecting back

This account of Torkild's first day of fieldwork in the Swedish sportswear industry raises several issues related to the challenges of doing ethnographic research, and it actualizes a number of bodily concerns that are involved in the everyday practice of fieldwork. Fieldwork is time consuming. Hence, the banal fact that it is physically impossible to occupy two places at one and the same time makes it difficult for full-time academics to bodily immerse ourselves in empirical field settings. In our experience, this is especially so as university employment has come to involve a range of different roles and responsibilities that compete for our energy, attention and participation. Even if you are able to conduct immersive fieldwork, it is difficult, if at all possible, to physically and cognitively participate in an activity whilst observing it and recording it. Let us share some of Torkild's reflections on this:

Whereas I had fully participated in the company's sports hour 10 days earlier, I found myself largely observing and taking notes during the morning meeting. Even during the lunch break I participated rather little in the various conversations that evolved. Whilst recognizing that my sheer presence implied some degree of involvement and participation, I was seeking to blend in without becoming completely invisible to research participants.

At the same time, I was anxious to build rapport with employees and gain their trust, realizing that the formal access I had been granted by management was no guarantee that I'd be able to do so. Rather, I was aware that empirical access is never infinite or unconditional, and I was prepared to negotiate my access to them on an individual, everyday and even case-to-case basis. To ingratiate myself with them, I was careful about how I dressed and presented myself, trying to come across as humble, friendly and palatable by not saying or doing anything that I thought they might be disturbed by or offended by (see, e.g., Ambjörnsson, 2004; Kertzer, 2009, p. 2).

Thinking back, this anxiety to fit in and be accepted probably made me excessively careful during the early stages of my fieldwork. Rather than championing the kind of 'rehearsed carelessness' which Garfinkel (1967) claimed crucial to pass as a member of another social category, this led me to micro-manage my bodily gestures, demeanour and utterances in such a way that I sometimes moved in an inhibited manner, gave stale responses when spoken to and bit my tongue instead of participating in the natural flow of conversation.

A particularly awkward incident occurred during one work-out session, when the temp trainer asked us to pair up and climb around each other. While I felt uneasy having someone else climbing around my waist (not simply because the added weight made it difficult to keep my balance), the inevitable intensity of physical contact made me more nervous that I was invading his intimate zone than he was invading mine.

Fortunately, I've become more relaxed over time – as I've come to know them better, as some of them have expressed interest in my study and in my presence, and as the increasing amount of material that I've generated has made me worry less that they may terminate my access. Thus, I've felt more comfortable trying to break the ice, initiate conversation and provoke unpredictable responses.

Despite my initial wavering, I ended up spending 18 months in the organization. Between September 2016 and February 2018, I participated in a number of different activities, including 35 sports hours and 53 lunch breaks, and attended 15 morning meetings and 17 workshops. I also had five meetings with the CEO and the HR Manager to present and get feedback on my observations and preliminary findings. These meetings proved a terrific opportunity to ask for expanded access, allowing me to later attend two job interviews, six staff appraisal meetings, three international sales meetings, one top management meeting and three fitness tests as well as shadow three middle managers for nearly a whole working day each.

Moreover, I've spent 50–60 unscheduled hours at head office, primarily by the reception and in the kitchen area. In addition, I've been spontaneously invited by staff members to join them in ping-pong games, and to sit in on apparel fittings. All in all, I spent approximately 310 hours at the company spread over 80 days.

At the same time, I wasn't able to participate as much as I wanted to or perhaps ought to have done; because of competing commitments, because meetings and activities were rescheduled at short notice, and because others, or I, had to cancel due to illness. Furthermore, I continually found myself negotiating my access to the company and to people in the company as well as my role in the field and my own embodied presence and absence. There are countless instances when staff have ignored me by taking detours or avoiding eye contact, when I've struggled to find appropriate words to make contact or initiate conversation, when I've sensed that they've been too busy or disinterested to talk to me, or when I've tried to join lunch conversations but been ignored and effectively put in my place as an outsider.

Nevertheless, attending and participating in these activities have enabled me to generate a fairly rich mass of empirical material, which hopefully elaborates, nuances and challenges some of the initial prejudices and preconceptions that I entertained before entering the site. In summary, the fieldwork made it possible to follow daily activities and work routines but also participate in a number of activities which management organized in order to break the monotony of daily routine.

It is worth saying a bit more about the contrast and connection between the mundane flow of everyday work and out-of-the-ordinary activities. As we shall see later on in this chapter, short-term shadowing enabled Torkild to learn what an ordinary work day might be like. In contrast, the sales meetings and the occasional after-work party and eating contest gave an opportunity to experience how less ordinary events are utilized to motivate employees struggling to realize their potential and reach their goals. These events actualize further challenges related to the embodied work of ethnographic field research, including evening presence, eating and drinking habits, and the management of first impressions. To illustrate, Torkild will share his experience from his first sales meeting.

Torkild's first sales meeting

It's 9 November 2016. I arrive at head office at 9.10am. The international sales meeting is due to start at 10am sharp. I feel tired and depressed. Nervous about the outcome of the US presidential election I woke up several times during the night. Earlier that same morning my worst fears were confirmed as news broke that Donald Trump had declared victory. But as I walk through the glass doors at head office people seem as content as ever, greeting each other as if nothing has happened. A couple of employees are even high-fiving. 'Don't they care? Or do they, despite all expert commentary, anticipate benefiting financially from Trump's politics?'

Well, if Goffman's (1959) theory of impression management has taught me anything it is to never judge a book by its cover. Rightly so. Only a couple of minutes later, a young staff member walks in, muttering that 'This is the end of the world.'

While I'm sitting at my usual spot in the reception area a steady stream of people start arriving. Probably employees and distributors from abroad. I get chatting with a couple of people, but when I realize that they take me for an employee I make sure to mention that I don't actually work here but am here to study their organizational culture. They're friendly and talkative, a bit older than the average employee at head office. They seem less careerist and more relaxed, though that might just be because they view the trip as an away-day. They don't strike me as fitness types, but remind me more of the local store managers I met the previous week. One man arrives on crutches. By 10 o'clock the reception and kitchen areas are packed with people chatting, having coffee and sparkling water, and helping themselves to smoothies and rye-bun sandwiches, while local staff are running around making final preparations for the formal programme which is just about to kick off in the showroom.

Avicii's 'Waiting for Love' starts playing at high volume, and someone tells us to take our seats. I join the crowd and manage to find a vacant seat next to a local staff member in one of the back rows. The showroom is filled to the brim. A number of local employees are standing by the wall. While the DJ is turning up Avicii to full blast, I manage to overhear a couple of people talking about Trump. They sound worried, and I'm quite moved by the whole situation: by the music, the lyrics and sitting so close to people in the jam-packed room. I sense hope, that there's a way out of the tragic circumstances caused by the US election.

At 10.35 the CEO walks up on stage, smartly dressed as usual. He reminds us of 22 June 1974, which he describes as a milestone in the life of the person whose name the company carries:

'That was a beautiful summer day', he states calmly. 'Unlike today. But he always aimed high, he did. He refused to change the goal ... And one thing is clear. We're not gonna change where we're going. Never ever!'

The CEO goes on to outline the corporate framework, blending KPIs, numbers and strategic questions with anecdotes and motivational speak. He talks non-stop for more than half an hour, the speed and volume of his voice increasing while he is addressing the audience. At 11.10 he reaches the end of his talk:

'You see, it comes down to one thing. How badly do you want it? Are you prepared to do what it takes? Now, let's turn the fall of 2017 into the best season we've ever done!'

To my puzzlement, the applause is tempered, and it is soon silenced by the music which again starts blasting out of the loudspeakers. The audience seems more enthusiastic when the otherwise mild-mannered Head of Marketing kicks off his afternoon talk. Asking everybody to stand up, raise their arms, and join him, he looks up and says slowly but loudly:

'Fuck you, Mister President!'

At 3pm it's time for another high-intensity sports hour, and by 4.30 most people stumble back, presumably worn out like myself, exhausted and with aching muscles, for after-work drinks and dinner. The DJ has left, but there's beer, wine, soft drinks and sparkling water, crisps and salted nuts, and later on, two or three different dinner options (meat, vegetarian and possibly something else). While these are ordinary ingredients at most after-work parties, I'm surprised to witness the consumption of alcohol here; at one of my first meetings with the CEO he had mentioned that he very rarely drinks alcohol:

'It doesn't add company value and it doesn't make people stronger.'

I'm carefully sipping on a beer with the Norwegian team when the CEO walks by, reluctantly handing out beers: 'I thought I might find you here', he says to me in a half-joking, half-disapproving tone, well aware that the Norwegian office prefers the pub to the gym.

Reflecting back again

We have already noted how ethnographic research requires a long-term commitment to field immersion, and in Torkild's experience this became increasingly apparent as he met new people, witnessed unexpected situations, and participated in a variety of activities. Even though he never reached a point when he could take his participants' trust for granted, it is likely that his field experience would have been much more restricted had he only spent time there for a few days or a week.

While this may have reduced his autoethnographic awareness of being a researcher and outsider in their space, it gave him a stronger sense of what it may be like to live and work in their culture:

> Torkild: Though incomplete, my increasing immersion affected my ability to observe them and their bodies, offering new opportunities for using my sensorial apparatus to explore their world, yet limiting my ability to do so. Over time I was subjected to more perceptual stimuli; to more sounds and smells, more bodily movements, gestures and facial expressions, more bodily actions and face-to-face meetings, more roars of joy and

sighs of exhaustion, more laughter and heavy panting, more handshakes and hugs, and more physical contact, with and without boxing gloves. Naturally, as I became increasingly busy conversing and participating, I became less able to devote my full attention to observe them.

Embodied immersion in a newspaper advertising department

The challenges of bodily immersion played out somewhat differently in David's fieldwork. When doing his PhD back in the 1970s David had gained access to a regional newspaper to study a sales department that was in dispute with its management after the company had brought in management consultants who recommended making one third of the staff redundant. At the time he arrived on the scene, the staff had just been reinstated as a result of their bringing the whole of the newspaper's labour force out on strike, which resulted in an immediate capitulation by management. While the story reflects a classic case of inadequate knowledge and the absence of any bodily engagement on the part of the consultants that recommended the redundancy programme, it is worth recording here not least because David's bodily presence and participation further actualizes the issue of trust in fieldwork and the management of boundaries between insiders and outsiders:

The consultants were ex-army officers who at best had a limited understanding of industrial relations and trade unions because they failed to investigate the staff in the sales department[1] where they were seeking to make serious rationalizations. Conventionally, sales staffs are not inclined to join unions, and this was the case here but only insofar as they were not members for purposes of negotiating terms and conditions with their own management. However, they were members of the same printers' union since this was a condition of their securing part-time highly remunerative part-time incomes from working at another newspaper on Saturday nights preparing the Sunday morning papers. Consequently, as soon as the redundancy programme had been announced the sales staff contacted the official union organizer and set up their own 'chapel' (branch) with an elected 'Father of the Chapel' (Shop Steward).

Meanwhile, the union as a whole for the company went on strike and closed down the presses although this was necessary for only one day as the management instantly reinstated the staff and abandoned the redundancy programme. Unfortunately for management, this was not the only cost to the company as the newly formed chapel were militantly aggressive in their refusal to cooperate with management, and this lasted for almost 10 years. In effect, management were denied their right to manage, which ordinarily would take such forms as staff training, internal competitions, accompanying the sales persons on visits to customers and calling customers as a way of monitoring the staff's activities. The result of all this was that the staff felt neither obliged nor had any incentive to work hard or even at all in many cases. This presented me with numerous research difficulties and, at one level, a problem that was impossible to resolve.

One of the main difficulties, which, despite every effort I failed to resolve, was the suspicion on the part of all these sales persons of my being a management spy. This was especially the case among those who had previously been earmarked for compulsory redundancy for they were convinced that I was preparing material for management to carry out their original plan to dismiss one third of the staff. It was those staff who eventually engineered my departure from the research site since they held their suspicions to the very end even though I made every effort to engage them with the research. My access to the research site had developed out of discussions with a personal friend who had worked in the department in the past and introduced me to his colleague and friend who continued to work there and with whom I made friendship before approaching the company for access. I had of course to seek access through senior management in the company and this I did through an interview with the Marketing Director who had responsibility for all the sales departments. Obviously in speaking to management I did not declare my friendship with one of the staff or my knowledge of the events leading up to the stand-off between the department and management. I simply suggested that this would be a good location and that I had prior experience of working in sales. I presume but cannot know whether or not the Marketing Director felt that the situation was so dire as to be virtually impossible to make worse by the presence of a university academic conducting a PhD, but the outcome was to secure full research access.

One of the methodological nightmares we have as empirical researchers is to confront participants who, despite our best efforts, refuse to engage with us:

> 'Look, I don't want to talk to you because if I tell you the truth it won't do me any good and if I give you a "line", it won't do you any good.' (Participant, Local Advertisers, Knights, PhD, 1984, Chapter 5, Preface).

My solution to these difficulties was to gain the trust of the Father of the Chapel – Jeff (pseudonym) – although this was made easier because of my friendship with one of his colleagues. However, shortly after entering the organization, my friend left to pursue a degree course in my own school at the university, but Jeff was fairly intellectual and was also tempted to return to education, so we had many enjoyable and embodied conversations. These all extended to most of the other sales representatives who, because of his endorsement, were prepared to give me the benefit of the doubt as to whether or not I was acting as a spy for management. However, the cost (if cost it could be called) was for me to participate in an elaborate and extensive communal set of hedonistic activities undertaken during working hours and only made possible by the inversion of power relations between management and staff within the workplace.

A typical day at the office, so to speak, in which I participated at least a couple of days a week for several months, would begin when the sales representatives would sit behind their desks with newspapers. Ostensibly and as a formal expectation, they were scanning the

(Continued)

(Continued)

pages to seek out opportunities for selling 'run of paper' advertising, but actually they were just reading whatever took their fancy. At 10am there was an agreed meeting in the canteen where, over their coffee or tea, a plan would be made for the day's leisure activities; these varied from trips to a seaside resort or other place of tourist attraction ('super-duper skives'), to arranging where to play golf ('super skives') or just spending most of the day first in the pub when it opened at 11am followed by the Press Club after the pub closed at 2pm ('skives'). They would be back in their offices just before they were due to finish for the day and management was oblivious of the virtual absence of any visits to customers to sell advertising space. Apart from the change in power relations, another reason why this could continue without bankrupting the organization was that the newspaper had a quasi-monopoly in the region such that there was a fairly strong demand to place adverts in the newspaper independently of any promotion by the sales staff. Also, classified ads provided a steady income, which alongside the copy price could keep the newspaper in profit.

Most of my time as a researcher was spent in the pubs, where I could talk to the reps in an embodied and often intimate manner although the heavy alcoholic drinking had both negative as well as positive effects. There is of course an association between heavy alcoholic drinking, skiving from work and masculinity,[2] but the culture of the printing industry generally and this department and organization in particular was highly masculine, so that was a considerable factor.

Having studied some social anthropology at university, I was fairly convinced that in order to get behind the formal presentations and official ideology of an organization, it was necessary to 'hang out' with the participants, and if this involved drinking withthem so be it. Negatively I must admit to fearing becoming an alcoholic during this process, and also there is no doubt that my note-taking was inversely proportional to the volume of alcohol consumed although I did try to fill in the gaps from these lapses the following day by excavating my memory. Moreover, while fully committed to empirical evidence even at this stage in my career, I was aware of the impossibility, if not the undesirability, of seeking accurate or 'perfect' representations of the activities under research investigation.

The negatives, in my view, were well compensated for by gaining understanding of the lives of these sales representatives as embodied subjects. Very little was hidden in the discussions and conversations that we engaged in as fully embodied friends talking and gesticulating about our lives. Clearly by virtue of participating in the range of skives, I had gained the trust of the staff and this was consolidated in particular through my engagement with Jeff. Nonetheless, even though since the formation of the chapel industrial relations had changed in the employees' favour, if management had evidence of the violations of working practices that I observed, it is unlikely that the official union would have defended them. This continued for several months and I soon began to realize that once the novelty of exploiting their employment wore off, it seemed collectively self-destructive. However, as Jeff would constantly remind me, he rationalized the militant stance he and his chapel took with regard to management:

> 'if we are not constantly fighting with management the men [*sic*] get restless and think nothing is happening.' (PhD, 1984, Chapter 5, p. 38)

One of my arguments in the thesis was that, in contrast to other contexts where union militancy is a function of strength (e.g. among UK car workers in the 1960s), in this sales department, the militancy was a function of weakness and vulnerability grounded in the anti-union predispositions of a majority of Jeff's members. Conventionally, sales people enhance their identity by increasing sales and income through bonuses or pay rises, but this was now impossible because of a refusal to cooperate with management. However, the one power that management retained here was control of the purse strings and no extra income would be forthcoming without evidence of productivity. This, combined with the sense of meaninglessness that comes from day after day of no other purpose than excessive and often 'debauched' leisure, resulted in everyone feeling frustrated, if not seriously depressed.

Detecting this and when writing up the material for my PhD, I sought to develop an analysis that provided a theoretical understanding of the stand-off between employees and management.[3] My analysis sought to draw on theoretical ideas about how power cannot really be exercised without some degree of responsibility and that the lack of this on the part of these sales representatives partly led to their unease. While management had lost much of their power except over financial matters, they still were responsible for the survival of the organization, which would have been in grave danger had it not been for its quasi-monopolistic market position. In having Jeff's trust and because he was increasingly showing interest in academic work, I gave him a copy of what I had written. In his belief that the analysis could help his staff retrieve a more realistic compromise with management that would facilitate negotiating for pay increases, he passed the report on to all the staff. The response of those who had previously been threatened with redundancy was to confirm their original suspicions that I had been brought in to spy on the staff and they felt that, if management saw the report, it would legitimize the original redundancy decision. Through the union, they instigated a vote of the department's staff to withdraw permission for me to continue the research and, despite Jeff campaigning for me to stay, I lost the vote by 11 to 10 and had to leave the organization. This was 18 months after the initial research access, but my Head of School suggested it was probably for the best since I might never have left of my own accord. It was several years later that the union compromised with management in ways not dissimilar from what I had been intimating, but management were never given sight of the paper for clearly the employee deviance would have provided instant grounds for dismissal.

To conclude, I have given the full details of this narrative to demonstrate how it would have been impossible to study these events in any way other than through an embodied engagement with the participants. Clearly, any other method of research such as formal interviewing or questionnaires would almost certainly have resulted in sufficient dissembling to conceal what was happening. However, I was unable to discuss matters with management because of the suspicion about my spying so I cannot say for certain what, if anything, they knew about this workplace deviance.

SITTING, MOVING, SHADOWING

To further understand the work that individual research participants do on an everyday basis, ethnographers in our own field of management and organization studies have stressed the need to stay physically near them for a substantial amount of time – for a full work day, an entire work week, or even longer. As Van Maanen (2011) points out, he has spent many hours sitting down next to several of the police officers that became part of his field studies, looking over their shoulder while they were working at their desks in order 'to (gulp) "get inside their heads"' and get a sense, albeit 'tentative and uncertain', of how they experience their 'work-a-day lives' (p. 227).

This technique is sometimes associated with 'shadowing'. Although Torkild's anthropologist friend Raoul Galli recently told him that he had never seen or heard the term used by any of his colleagues, over the past few years shadowing has become an increasingly popular ingredient in organizational field research. According to our colleague Barbara Czarniawska (2007), who may have written the first systematic treatise on the subject, shadowing involves following someone as they go about their everyday business.

Czarniawska has located the first explicit reference to shadowing in the Italian sociologist Marianella Sclavi's (1989) comparative study of US and Italian schools. Taking inspiration from Truman Capote's (1980) short story 'A Day's Work', where Capote followed a cleaning woman around work for an entire day, Sclavi followed the neighbour's teenage daughter to school for two weeks.[4] Shadowing, then, is typically not sedentary. Shadowing requires researchers to follow research participants around, to sit when they sit, to move around with them when they are on the move.

Becoming a shadow?

At the first feedback meeting about two months into his fieldwork, Torkild asked the HR Manager if he might be allowed to shadow two or three employees over the course of a few hours, a work day or a few work days: 'The idea is to get a more intense experience of what it's like to work here.' The HR Manager liked the idea and promised to check with a few staff members who might be comfortable with having someone follow them around for a day. A few days later, she had set him up with three middle managers in three different departments. The deal was that Torkild would follow each of them around for a full work day.

When I get ready for my first shadowing stint on 29 November I'm both excited and a bit nervous about this opportunity, particularly since I've heard rumours that one of the leading scholars in the area ended up having her shadowing cut short because of poor chemistry with her shadowee.

At 10.58am I walk upstairs to meet the sales manager who I will be shadowing for the rest of the day. While we've talked briefly during lunch and coffee breaks, I haven't had any long conversations with her and I know little about her work. However, I know that she joined the company before the new CEO. Thus, I'm optimistic that she can help me better understand how the company has changed and how the new performance culture has affected her, with its focus on key performance indicators, 'smart' work practices, measurable goals and physical exercise. I'm greeted cordially as I enter the office landscape:

'As you probably all know, this is Torkild. He will be my shadow today, to see what kind of work we do around here.'

I've met most of them before. While her colleagues briefly look up from their screens and keyboards to offer a friendly 'hello' or 'good morning', she tells me to pull up a chair and sit next to her so that I can get a better view of what she is doing.

'Don't worry', she says when I reassure her that I'll try and disrupt as little as possible. 'No one else is using that desk anyway today ... And please feel free to ask questions. I'll do my best to explain.'

Before I manage to think of a suitable question she pedagogically explains what the different people in the department do and how she is just in the middle of checking stock prices and quantities and adding extra pieces from next year's collection to incoming orders:

'You see, our wholesalers are starting to make extra orders, and then we just realized that three cases have been stolen from our old warehouse.'

While she is filling me in on excess purchase estimates and stock balancing I ask her what she enjoys most about her job. She looks up, smiling:

'The purchasing bit. Purchasing new collections for our retailers ... the way we sell off old stock is a bit chaotic. There are different codes for each piece, but our customers place their orders in blocks of pieces ...'

Not sure that I fully understand these technicalities, I feel I'm on safer ground when she starts talking about her managerial role, about how she is responsible for new staff members in the retail department, their performance appraisals and so on.

The atmosphere in the small office area is calm and friendly. People make business calls, process tasks on their computers, and help each other out with queries about computing issues, customers and contact persons. They seem focused on their work, when one sales assistant in his mid-20s suddenly addresses me:

'Did you get this, Mr T?'

'No, I'm afraid not', I stutter, a bit perplexed.

'The massage. Whenever you want something from B you have to rub her shoulders.' 'It's so great', B adds, as I silently wonder if her aching shoulders are caused by her desk work.

They're all processing stuff on their computers when the same sales assistant interrupts again to ask my shadowee L if she feels as if being under surveillance:

'Well, a bit, though now I'm just doing this you know.'

Before I can think of a good way to ask them how my presence affects their personal integrity, it's almost lunchtime. L wants to sort out something with Logistics but is persuaded to leave it till after lunch. L, B and I go downstairs to meet up with a male colleague in Marketing. They've booked a table at the Korean restaurant around the corner to help B plan her baby shower. I tag along but find it difficult to participate in their conversation. Work is only briefly touched upon, and I have no children, so it's not as if I can share my own experience on the subject matter.

(Continued)

(Continued)

When we get back to head office a group of people from Production are playing ping pong. 'Sorry, but we don't have time', L and B explain. I pop to the loo, and when I get back L is on the phone upstairs with a recruiter she is due to meet in half an hour. I find my chair and start jotting down notes from our lunch. L first returns to her spreadsheet, then to her emails. She takes a deep breath and seems a bit frustrated that those last items haven't been priced yet – in other words, that parts of her job depend on other things running smoothly, on other people doing their job, on time. A few minutes later, H from Logistics comes by, and asks L in an upbeat voice: 'L, you wanted help with those prices, didn't you?' 'Yes please', she answers, in a half-frustrated, half-relieved manner. As we walk downstairs, H jokes that L is extra nice today since I'm shadowing her. The pricing issue is sorted out in five or ten minutes, and I follow L to her next meeting with the recruiter:

'He's a friend of a friend, we met at their wedding, so I just couldn't say no.'

L introduces me to the recruiter, a slim and smartly dressed man in his early 30s. They catch up for a few minutes before they get down to business. The recruiter is eager to get them on board his company's digital recruitment platform. I suspect he would do his sales pitch non-stop if given the chance:

'We have a large pool of relevant job applicants', he explains, 'and you can view their videotaped interviews before you decide who you want to interview face-to-face ... You lease access to our pool on a monthly basis.'

They talk for 20 minutes or so, not minding me much, and L seems energized by this break of routine. She tells him to get in touch with the HR Manager and encourages him to get something from the shop before he leaves.

My wristwatch shows 2.30pm. L and I get coffee and climb the stairs to the second floor. We're both tired. She opens the spreadsheet on her computer, and continues balancing stock and processing purchasing orders.

'We're moving warehouse, you see, and that's creating quite a bit of extra work at the moment.'

L checks her email again, efficiently churning out 10 replies in 20 minutes. 'What an email harangue!' she exclaims, and I can't help but be impressed by her tempo and how she's handling a range of different issues almost without breaks.

At the same time, I'm doing my best to observe what different people are doing and listen in on their conversations. The volume is higher than before lunch. Staff from other departments are coming and going, people are speaking on the phone, asking each other for help and advice, interacting to solve problems, teasing each other and exchanging banter about difficult customers. By 3.30pm they look rather knackered though. We all are. I imagine that the low ceiling is preventing enough oxygen to get into the room.

'Traptraptrap!' L is impatiently drumming her nails against the desk while calling the helpline at one of their software vendors. Getting no answer, she looks at tomorrow's schedule. Smiling, she asserts that she has no meetings scheduled:

'Hopefully, that means I'll be able to sort out what I couldn't finish today. Then, on Thursday I've got a meeting with Logistics about coordinating our system with our haulage contractor.

On Friday we've got our Performance Management Day, and then next week I'll be making purchasing estimates with my boss.'

As we're both about to head off home, she hands me a copy of a sentimental bestseller which was given to all employees the previous week:

'No, you have it', she insists when I bashfully try to decline her gift. 'I'm never gonna read it anyway.'

Naturally, each of Torkild's three shadowing sessions yielded unique experiences and insights into work and life at the sportswear company. As he was allowed to sit close to his shadowees and see what they were doing he developed quite a detailed picture of what an ordinary work day may look like for each of them, especially with regard to their work tasks, routines and responsibilities. However, it also turned out that they had a number of things in common; in terms of how they experienced work, how they interacted with managers and colleagues, and how they related to organizational structures, systems, norms and values:

> While my shadowees seemed to carry on with their work as normal despite my presence, they took time to explain various parts of the job to me, especially key tasks and managerial responsibilities. All three of them entertained cordial relations with colleagues and managers, and even though I witnessed a fair amount of joking, teasing and bantering within their teams, it was clear to me that they were there to work; they were determined to get the job done and reach their targets. There was little time for breaks between meetings and other work tasks, and they all seemed to be getting tired as the work day was coming to an end, as did I. Indeed, during the days that I shadowed them none of them seemed to adopt the various time-management techniques endorsed by management but carried on with their tasks until they were completed.

At the same time, two incidents on Torkild's second and third days of shadowing gave rise to two delicate ethical issues which we, for integrity reasons, have decided only to mention briefly here:

> On my second day of shadowing, I sat in on a number of appraisal meetings where the team leader had already filled me in with details about each team member. Concerned that this made them extra vulnerable, I took greater care than usual to assume a humble bodily manner, though it's difficult to know how apparent this became through my posture, gestures and facial expressions.
>
> On my third day of shadowing, I attended an underwear fitting with one of the design teams, the purpose of a fitting being to see how a prototype garment looks on a real person's body and check what needs adjusting before it goes to production. As the

> fitting was performed with a live woman model, I was acutely aware of being the only male in the room. To avoid imposing a male gaze I made sure to look away or look down in my notebook as much as possible during the half hour which the fitting lasted.

It is not obvious, however, that Torkild's behaviour in these two situations was appropriate, or that it had the intended effect. Our point is that both incidents were face-to-face encounters which prompted a laboured embodied response.

BODILY CONSCIOUSNESS, AFFINITY AND INTERVENTION

While there is necessarily a participative, embodied and autoethnographic aspect to all ethnographic fieldwork, one central question which emerges out of these fieldwork accounts is how much we have been able to bodily immerse ourselves in the field and what issues this has raised. To what extent has Torkild been able to get a sense of what it is like for ordinary employees to live and work in a performance culture? And how well was the young David able to embody the culture of the newspaper advertisers? What bodily practices did they enact in order to become part of these cultures, and what physical, ethical and political constraints prevented them from doing so as much as they wanted to? Since it remains quite fresh in his memory, let us focus on Torkild's case here.

Through his accounts above, it becomes apparent that Torkild's presence, participation and immersion was profoundly and consciously embodied in a sense which goes beyond the physical exercise he did during the company's sports hour. Quite similar to Shusterman's (2005) notion of somatic consciousness (discussed in Chapter 2), these conscious bodily practices were things that Torkild did in order to fit in, avoid upsetting participants and maintain access. Particularly in the early days of fieldwork he worked hard to manage his body language and tone down his bodily gestures. He was careful about what he wore to the site and what food he had for lunch, and he did his best to suppress tics and other irritating habits.

There were other constraints to his immersion too. During the first few months Torkild's knee injury and inferior fitness level limited his participation in certain physical activities. However, over time, as his general physique improved, he became able and eager to go to as many sports hours as possible:

> Despite my secure employment as a university professor meaning that I was free to leave the site if things got tough, I believe that I got at least some sense of what it's like for different employees to live and work in a fitness-driven performance culture. While the increasing pressure to publish in competitive journals, get media coverage and be liked by students has turned academic life into a performance culture on a par with many

performance cultures in the corporate world, I've recognized that my performance pressures are similar if less direct and penetrating than those facing my participants. And by exercising with them in high-intensity work-outs I've certainly felt an affinity with their physical pain and exhaustion, the increase in adrenalin and endorphin, and their frustration over injuries that strike when they've pushed themselves too hard.

Torkild's participation and immersion has affected him, then, even in an embodied sense. But to what extent and in what ways has his bodily presence affected them? Although Torkild has not had the opportunity to ask his research participants, the so-called Hawthorne Effect, which was induced from experiments conducted between 1927 and 1932, may shed some light on this. The Hawthorne Effect has been subject to a number of conflicting interpretations (see, e.g., Mayo, 1945; Landsberger, 1958; Braverman, 1974), but one possible interpretation is that people are likely to modify their behaviour according to prevailing social norms when someone deemed to be of importance shows interest in their work. According to Mayo's (1945) early reading, the women workers in the relay assembly room at Western Electric's Hawthorne Works increased productivity while being observed by a team of researchers. We may add that this was probably reinforced by the fact that they were observed by a team of male Ivy League professors wearing white coats.

Even though the possible status associated with academics in corporate field settings are likely to have decreased during the several decades that have passed since the original Hawthorne experiments, Torkild's presence at the sportswear company and his participation in activities endorsed by management may have added weight to these activities, making employees view them as more important than they otherwise would have done. Thus, rather than disrupting the setting and making management and employees think about what they were doing, Torkild may have ended up reinforcing management's agenda, adding pressure on employees to become fitter and optimize their performance:

Indeed, management seemed keener on my physical presence at the sports hour and at various staff training sessions than to hear about the preliminary findings I presented at our regular feedback meetings. After a couple of months, they started referring to me as 'their professor' and they became increasingly positive about my presence as I started to become fitter and lose weight. Perhaps they viewed my bodily progress as a sign that I appreciated their fitness regime, that they'd convinced me that they did the right thing, and that I eventually started to become more like them?

This kind of relationship is not unusual, and it no doubt strikes a chord with David's long experience of doing fieldwork in organizations where empirical access is granted by senior managers:

Fairly early in my research career I had negotiated complete access for myself and for five of my colleagues to do fieldwork in a financial services company. I knew the business having

> worked in it for several years before becoming an academic but I also knew the CEO via my brother who was a CEO of another company. Anyway, we were 'given carte blanche to interview at any level in the organization, attend and tape record committee proceedings (including the equivalent of board meetings), visit branches, read documents and sit in on training sessions and corporate conferences and rallies' (Knights and Murray, 1994, p. xiii). But when I offered to give the CEO a report of the research when it had been finished, he responded: 'David you don't need to do that – it is sufficient that you are just here because your presence and questions to staff will force them to reflect on how they do things and that should stimulate them to work better and be more productive.'[5]

This is not to deny that the practice of giving feedback on observations and findings to key contacts, participants and gatekeepers may be a contentious process, particularly in fields where power is dispersed and when your empirical access is decided by different stakeholders or groups. Whereas early reporting sessions with management may help clarify questions, generate interesting responses and open up new opportunities for data generation, various actors may have their own concerns about what is being reported and when. In Torkild's case, management at the sportswear company had first seemed keen on him doing a plenary presentation with head-office staff at the end of his project, but they later ignored his proposal to do one, possibly because they did not want all staff to take part in his more critical findings. And as we saw earlier in this chapter, David's feedback session 18 months into his PhD research project in a UK regional newspaper ended up denying him further empirical access.

SUMMARY: EMBODIED IMMERSION AND REFLECTION

In summary, ethnographic fieldwork actualizes a number of bodily issues and questions, which makes it suitable for pursuing an embodied research process. As it usually requires researchers to be physically present in a specific field setting over a long period of time, it offers unique opportunities for researchers to participate in and immerse ourselves in the settings we study.

Field immersion will generate rich material about many aspects of research participants' lives and bodily practices – about their gestures, facial expressions, mannerisms, physiques, everyday habits and face-to-face interactions. Meanwhile, you will be relying on your own body to participate in their lives and their cultures – using your body language, gestures, facial expressions and basic habits and movements when interacting with them, talking with them and getting to know them. Whilst you will be observing their bodies in order to understand how they relate to themselves, to other people and to sociocultural norms and structures, they will be observing your body, seeing how you relate to them, through your body.

You are likely to feel things too. For one thing, you are likely to experience things which are related to your role as a researcher, whether discomfort at being an outsider peeking in, excitement at 'juicy data', or stress from the conflicting pressures of spending enough time in the field and on campus. Insofar as you have started to live like your participants, you may even feel and experience some of the things they feel. Moreover, your physical presence is likely to generate feelings and sensations amongst *them*. Some participants may be uncomfortable with your presence. Others may feel aggrandized by the attention you show them.

We shall return to discuss some of these issues in later chapters. But let us first turn to the role of interviews and conversations in embodied research. As we have shown in this chapter, immersing yourself in field settings offers opportunities to talk to people and have conversations with them, about trivial things and about life and work in general, but also about things that are specifically related to the topic of your research. In the next chapter we shall therefore look more closely at the use of conversations in embodied research, from chats and informal conversations to interviews, and discuss what particular challenges they pose.

Questions and exercises

1. What makes ethnographic fieldwork suitable for embodying the research process?
2. Presume that you are planning to do fieldwork in a community, an organization or with a specific group of people. What bodily senses would you be relying on when doing so? How? Why? What bodily features and expressions would you watch out for when observing participants?
3. What kinds of dilemmas do you associate with ethnographic fieldwork? What challenges may ethnographic fieldwork pose not only to the bodies of researchers but also to their health and wellbeing?
4. Do a mini-ethnographic study where you observe people at a particular site, and, if possible, participate in what they do. Write down your main observations and reflections. What was particularly difficult about this task? To what extent did you experience a trade-off between observing their culture and participating in it? In what ways did this difficulty play out during your study? What dilemmas emerged as you conducted this study? Looking back, what would you have done differently?
5. Shadow a friend or a fellow student for a day or for a few hours. Write down your main observations and reflections. What was particularly difficult about this task? What dilemmas were you faced with as you conducted it? What would you change if you were to conduct this task again?

5
EMBODYING INTERVIEWS AND CONVERSATIONS

FROM WORDS AND FACTS TO LIVED EXPERIENCES AND EXPRESSIONS

Most methodology books and empirical studies that subscribe to interviews and conversations as means of data generation tend to view these methods as sources of verbal information, facts and textual content. However, when interviews and conversations happen face-to-face between researchers and participants, they are embodied encounters where experiences, perceptions and cultures are expressed. The views and ideas communicated may prompt and be shaped by a range of bodily actions, practices and feelings, including gestures and facial expressions, appearances and accents, nervousness and enthusiasm, energy and exhaustion. They may also be affected by physical dimensions such as proximity and distance. Moreover, our bodies may communicate meanings which affect the power dynamics between researchers and participants, particularly as they are inscribed by identify markers such as gender, race and class.

In this chapter we will engage with some of the embodied aspects of interviews and conversations and discuss how this affects the work we do when interviewing and talking with research participants. But before doing so we need to say something about the disembodied assumptions, principles and practices which dominate existing writings on the research interview as a form of empirical enquiry.

EFFICIENCY, DISTANCE AND FAST FACTS IN 'LONG INTERVIEWS'

One of the most popular and widely cited treatments of the interview process is Grant McCracken's (1988) short book *The Long Interview* (with more than 7,000 citations on

Google Scholar as per 9 May 2018). Whereas the embodied research practices that we are seeking to foster through our book are geared towards crafting rich and lively accounts, McCracken's version of qualitative enquiry is aimed at structured, efficient and unobtrusive exhaustion of information. Actually, McCracken's approach is explicitly at odds with many of the methods that we promote here, including participant observation and what he calls 'the unstructured "ethnographic" interview' (p. 7). In contrast to these methods, McCracken insists that the long interview 'is a sharply focused, rapid, highly intensive interview process that seeks to diminish ... indeterminacy and redundancy' (p. 7). By design, then, the long interview is 'deliberately less obtrusive' and offers 'a highly efficient, productive, "stream-lined" instrument of inquiry' (p. 7).

According to McCracken, this makes questionnaires indispensable in long interviews. While their widespread use in quantitative studies has set questionnaires apart from qualitative interviews, they are at the centre of McCracken's interview paradigm because they allegedly 'ensure that the investigator covers all the terrain in the same order for each respondent' (p. 24). We have little doubt that the structured use of questionnaires may help remove indeterminacy from the interview process and reduce the likelihood that interviewers will come too close to their interviewees. Indeed, McCracken puts particular emphasis on the need to manufacture critical distance in long interviews, adding that this is more likely to be achieved when studying other cultures where everything is likely to be new and 'mysterious' to you (p. 22).

However, McCracken's paradigm inadvertently raises several questions about the nature of qualitative research and what purpose it serves. Even if we accept his assumption that long structured interviews and the social relations they involve can be managed to the extent that he indicates that they can, what are the consequences of his emphasis on structure, focus and streamlining, efficiency and rapidity, consistency, intensity and determinacy?

More specifically, can we go beyond our own prejudices, and be surprised, if we keep too much distance? Are interviewees likely to open up and talk if they need to stick to a set sequence of questions? What if they rather want to talk about something else, which you may assume is redundant and irrelevant at the outset, but which may later turn out to be relevant, or open up new and more interesting lines of enquiry further on? Will there be time to build trust and open up lines of interesting conversation if the interviewer is mostly concerned with covering a particular ground and ticking off questions one at a time? What about all the other things that happen in interviews – things that may not be expressed verbally but expressed and experienced nonetheless? What about seating arrangements and bodily postures, looks and glances, smiles and frowns, cries and laughter, handshakes, high-fives and open palms, accents, pauses and silences?

In our view, McCracken's notion of the long interview does not merely ignore that face-to-face interviews are embodied encounters. He even ignores that they are social events (Seale, 1998) between people, and that they constitute affective encounters (Pink,

2015) where people exchange emotions and experiences as well as views and ideas through bodily and verbal expressions. In stark contrast to McCracken's formula are the real interviews that Eva Tutchell and John Edmonds (2015) did for their book *Man-Made: Why So Few Women Are in Positions of Power.* Here, Tutchell and Edmonds did long face-to-face interviews with a number of women who had successful careers as managers, public servants and politicians. The overarching aim of their interviews was to understand how they had achieved success within 'this man-made society' (p. xvi). Rather than relying on ready-made scripts and questionnaires, Tutchell and Edmonds sought to nurture an open conversation with their participants about their life and work experiences.

Interestingly, they only asked one direct question: 'Do you regard yourself as a feminist?' (p. xvi). Initially, this either generated polished spiels or short evasive responses, especially in the interviews that John Edmonds conducted. It was only when he made himself 'very quiet and very small, neither of which was easy' (he has told Torkild) that they opened up. Although Tutchell and Edmonds did not have very specific ideas about what they wanted their interviewees to open up about, this tack shifted the conversation from talk about career successes to talk about painful experiences: of sexual discrimination, harassment and even abuse.

Quite the opposite of McCracken's efficient search for fast facts, Tutchell and Edmonds' experience strikes a chord with the practice of ethnographic interviews, where researchers form 'respectful, on-going relationships with their interviewees', which allows for 'a genuine exchange of views and enough time and openness ... to explore ... the meanings they place on events in their worlds' (Heyl, 2001, p. 367).

HOW-QUESTIONS, WHY-QUESTIONS AND RICH RESPONSES IN FACE-TO-FACE INTERVIEWS AND CONVERSATIONS

By stressing that there is more to interviews than extracting verbal content from asking and answering questions, we are not suggesting that what questions we ask in interviews is immaterial. Since we conduct interviews and conversations in an effort to find answers to research questions, the questions we ask when interviewing and talking to research participants are typically shaped by the overarching questions that inform our research projects.

When Torkild interviewed consultants in a Big Four accounting firm in order to understand how managers and employees seek to manage wellness, he was far from immune to the popularity of McCracken's notion of the long, structured interview. Thus, he tried to ask all his interviewees a number of the same direct questions: 'How would you describe the company and the culture?' 'What do you most enjoy about your job?' 'What do you most dislike?'

> However, I soon found myself unable to ask all of my interviewees all the same questions, and it was virtually impossible to ask them in the same order. Trying to cover everything became hopeless, especially as some people spoke a lot more to some questions than others did, and it felt counterproductive to interrupt a person just so I could ask my next question on the list. Simply asking someone to tell me about themselves and their job often gave a far more elaborate response than some of the more pointed questions I had prepared.

In David's research with Caroline Clarke on the professional careers of academics and vets, they asked a variety of questions, some specific but others far more open-ended:

> We would for instance ask our respondents to comment on significant changes in the higher education sector over the last decade, but we would also ask them to describe a good day at work and a bad day at work. However, we would often open an interview with the following question: 'Imagine you are at a dinner party, and somebody asks you what you do, what would you tend to say?' Then we'd try and ask them how central they thought their work is to their sense of themselves.

In our view, it is these kinds of questions that facilitate a genuine conversation, as opposed to the mechanistic set of responses that are typically secured through questionnaires. One gets more value from interviewing if one is able to make it as relaxed and open as if it were a meeting over a drink with a friend.

Figuring out what questions to ask in an interview or a conversation is not only a matter of content. The ways we ask questions may have a big impact on the interview too: on how participants experience the situation, and on what they decide to share. Howard Becker (1998) provides an interesting take on this in his unconventional methods book *Tricks of the Trade*. Here, Becker argues that it is usually more fruitful to ask 'how' questions than 'why' questions. Asking people 'why they did something', Becker recollects, 'inevitably provoked a defensive response', as if the question was 'a request for justification'. Even when Becker asked them about their choice of career or education, his 'why' questions tended to be met with brief, guarded and pugnant answers 'as if to say, "OK, buddy, that good enough for you?"' (p. 58).

Realizing that he had to change his tactic, Becker started asking 'how' questions, such as 'How did you happen to go into that line of work?' and 'How did you end up teaching at that school?' According to Becker, these questions worked much better, producing lengthy answers and more elaborate stories which included details about how other people's actions had affected the choices made by his interviewees. Even when asking marijuana users how they happened to 'start smoking grass' (p. 58), none of them gave the kind of 'guilty reaction' that had been provoked by the more inquisitorial question 'Why do you smoke dope?' (p. 59).

While the effect that Becker's 'how' questions had on his respondents is well illustrated through interviews that he conducted with marijuana smokers around 1950 (Becker,

1973; see also Becker, 1953), we are equally intrigued by the rather embodied responses which this generated regarding the process and experience of learning to become a marijuana user. Particularly, one of Becker's interviewees speaks vividly about how 'flipped' he got even after 'four pokes'; how his 'heart was pounding real hard', how '[he] was sweating' with his 'coat open' even though 'it was five below zero', and how '[he] walked ... two blocks ... and fainted behind a bush' (1973, p. 57).

Even though we would mostly strive to set up interview situations which enable interviewees to speak freely to a topic, the social, political and embodied aspects of interviews which we are highlighting here make them rather different from the ideal speech situations that have been highlighted by Jürgen Habermas (1984 [1981]) and his followers. As Becker indicates, interviews can quickly turn into situations that the interviewee experiences as invasive and insensitive. Beguiling ourselves that we come to an interview with entirely innocent intentions may matter little to our respondents.

However, the power dynamics that are played out in interviews do not necessarily sediment established power structures between allegedly powerful academics and powerless participants. In our experience, the status attributed to academic researchers nowadays is somewhat different from the experiences that Becker draws from the early 1950s. Interviews may constitute events of resistance, offering opportunities for people to express dissenting views and experiences, and, to paraphrase bell hooks (1986), 'talk back'. This involves 'speaking as an equal to an authority figure, ... daring to disagree and ... having an opinion', even though that is 'to invite punishment, the back-hand lick, the slap across the face that would catch you unaware, or the feel of switches stinging your arms and legs' (p. 123).

Talking back in research interviews may take various forms; it would be a mistake to assume that it provokes physical punishment or that the researcher automatically holds a position of authority. As interviewers, researchers often find themselves in a minor position. Torkild had this similar experience when interviewing trans people as part of a study on transvestites' gender practices in everyday life and work (see Thanem and Wallenberg, 2016):

> One participant in particular was adamant to demonstrate how much s/he knew about the topic, especially that s/he was more knowledgeable than me. S/he even insisted that I ought to reorient the project to focus on the heteronormativity of cis people rather than the embodied practices and experiences of trans people. Even back then, I couldn't help but notice the irony in this, as this person relentlessly sought to help other members of the support group pass as 'real women' by instructing them on what to wear, how much make-up to apply, and what to eat and not. For fear of losing access, though, I never dared to bring this up.

Interviews, then, actualize a variety of bodily issues. In the interest of developing more embodied accounts from research interviews and conversations, let us now move on to say something about how such face-to-face encounters may be filled with non-verbal as well as verbal cues and expressions.

FACIAL EXPRESSIONS, GESTURES AND BODY LANGUAGE

As we have suggested above, interviews and conversations may generate rich accounts of people's bodily sensations, felt experiences and embodied practices. But insofar as interviews and conversations are conducted face-to-face, they are even sources of non-verbal communication. Facial expressions, bodily postures and physical gestures may be open to a variety of interpretations, which may complement and contradict what we actually say.

Before we try and illustrate how body language may add meaning in interviews and conversations, let us take a look at some of the basic ideas about body language and facial expressions. While the meaning and management of body language is given much attention in self-help books, business magazines and other popular media outlets, it is also a thriving research area in psychology, and it has a long if spurious history in philosophy and natural science.

In the treatise on physiognomy, which was written around 300 BC by a member of Aristotle's school, it was claimed that 'facial expressions ... accompany different conditions of mind, such as anger, fear, erotic excitement, and all the other passions' (Loveday and Forster, 1913, 805a25–30). Two millennia later, Charles Darwin (1872) did much to increase interest in the topic through his book *The Expression of the Emotions in Man and Animals*. Drawing heavily on the neurological research of Sir Charles Bell (1844) and Duchenne de Bolougne (1862), and working closely with the psychiatrist James Crichton-Browne, Darwin distinguished his project from the physiognomic interest in recognizing a person's character in permanent physical features. Instead, he set out to connect human and animal states of mind with how we move our bodies and bodily features.

Whilst admitting how 'difficult [it is] to determine this with certainty' (1872, p. 18),[1] Darwin develops his view around the principle of 'antithesis', namely the notion that opposite states of mind, such as aggression and caress, are accompanied by opposite physical expressions. According to Darwin, this applies to humans as well as to animals: while an aggressive dog 'walks upright and very stiffly' (p. 50), a caressing dog will assume a 'downward movement' (p. 51); and while a person who passionately would like to be left alone 'generally moves his arm as if to push ... [others] away', people who 'eagerly desire' another's company 'act as if pulling' the other towards them (1872, p. 6).

The book contains several examples and illustrations of how various feelings and emotions are expressed through the body. Darwin claims that this is most obvious in 'states of low spirits' such as grief and 'states of high spirits' such as joy: whereas grief is expressed through 'oblique eyebrows', joy is expressed through laughter. Further on, he argues that ill temper is expressed through frowning and 'firm closure of the mouth', that hatred and anger are expressed through sneering, and that surprise, astonishment, fear and horror are expressed through 'elevation of the eyebrows', through 'opening of the mouth' and through 'protrusion of the lips'. On Darwin's theory, these bodily expressions of emotion are innate and universal because they are found across different cultural contexts as well as among infants. At the same time, he rejects that the mind has much control over the body: as very few expressions can be governed by will, most bodily expressions are an outcome of habit or spurred by the nervous system.

Darwin's methods were mixed, and included questionnaires (possibly inspired by his cousin Francis Galton), simple experiments, literary illustrations (from Homer and Shakespeare) as well as observations of infants, mental health patients, people from different cultural groups and works of art. He also paid much attention to 'the action of galvanism' (1872, p. 18), that is, the effect which administered electroshocks would have upon a person's facial expression. The findings that Darwin reports on electroshocks were drawn from Duchenne's experiments on mental health patients. Duchenne (1862) had documented these effects through photography, and kindly let Darwin 'copy as many ... photographs as I desired' for use in the book (1872, p. 5).

Darwin's method has been critiqued for being unsystematic and unscientific. His questionnaire (1872, pp. 15–16) was distributed to a non-representative sample which only generated 36 aggregate responses, many of which were obtained from missionaries who had observed aboriginal people whilst having limited understanding of what meaning the individuals themselves had attributed to their own facial expressions. His experiments, which largely involved showing photographs to friends and family and asking them what feeling they reckoned had spurred a specific facial expression, suffered from a similar problem. Rather than comparing these responses with feelings reported by the people in the photographs, Darwin based his conclusions on 'consensual attribution' by his respondents; that is, he inferred that a feeling that was consensually attributed to a facial expression by his respondents was also a universally truthful feeling. Finally, it has been argued that Darwin merely offered illustrations and common sense explanations of how a particular feeling or state of mind was expressed. For instance, he simply assumed that a dog was happy when it wagged its tale without actually verifying that happiness was what the dog was feeling (see Russell and Fernández-Dols, 1997, p. 7).

Nevertheless, Darwin has been a significant source of inspiration for modern research on facial expressions and body language in psychology. One of the most influential scholars in this area is Paul Ekman, who has played a leading role in re-establishing Darwin's idea that facial expressions are universal. Complementing Darwin's method of consensual attribution with elaborate coding schemes and techniques for measuring facial muscular movements (see, e.g., Ekman and Friesen, 1976, 1978), Ekman (e.g. 1992, 1993, 2003) argues that human beings express a limited number of 'basic emotions'.

Basic emotions include those that we tend to enjoy feeling (such as amusement and contentment) as well as those that we generally do not enjoy feeling (such as sadness, anger and fear). Although Ekman and his colleagues acknowledge that the face is 'a confusing source of information' (Ekman et al., 1972, p. 1), that not all of our emotions (such as guilt, shame and embarrassment) are basic, and that basic emotions are displayed through 'families' of facial expressions which vary in intensity, he emphasizes that basic emotions have a number of common characteristics: basic emotions are distinct, universal and coherent; they are quick in onset and brief in duration; they are automatically appraised by those who observe them; and they are unbidden in occurrence (Ekman, 1993).

Building on these insights, more recent research has identified some 20 emotions that may be associated with more or less distinct facial expressions. For instance, pride is associated with tight eyelids, smiling and keeping one's head back; shame is associated with furrowed eyebrows, a raised lower lip and keeping one's head down; and sympathy is associated with parted lips, a dropped jaw and leaning one's head forward (Cordaro et al., 2018).

In his book *Emotions Revealed*, Ekman (2003) combines photographs and verbal instructions to demonstrate a number of bodily exercises aimed to help readers imitate basic emotions and recognize these emotions in ourselves and in others. To practise the facial expression of sadness, Ekman encourages readers to do as follows:

> Drop your mouth open; Pull the corners of your lips down ... raise your cheeks, as if you are squinting ... [and] Let your eyes look downward and your upper eyelids droop. (p. 95)

To practise the facial expression of anger, readers are given the following instruction:

> Pull your eyebrows down and together; ... open your eyes wide so that your upper eyelids push against your lowered eyebrows, staring hard. ... Press your lips together tightly and tense your lips; don't pucker, just press. (pp. 134–5)

When Torkild tried to do these exercises at home it produced the following results (Photos 5.1a and 5.1b):

(a) (b)

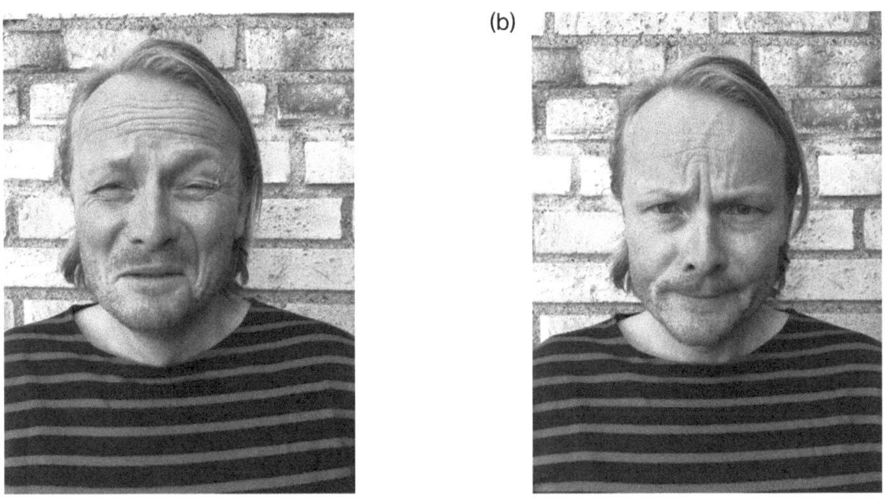

Photo 5.1 Photographs by Louise Wallenberg.

This does not mean that embodied researchers must learn to become experts at reading and deciphering other people's gestures and facial expressions so that we can know the complete truth about what they are communicating without the use of words. Many of the correlates that this branch of psychology asserts between emotions and facial expressions may seem self-evident. Moreover, if we accept the premise that many facial expressions of emotions are universal across different cultures and acknowledge that we observe and readily understand most of the facial expressions that people around us express on an everyday basis, we are such experts already.

But more importantly, facial expressions and other kinds of body language tend to be far more complex and ambiguous than Ekman's theory implies. While hand gestures may enhance what we think and say (Goldin-Meadow and McNeill, 1999; Beattie, 2003) or reveal mismatches between what we think and what we say (Cohen et al., 2010), recent research in psychology suggests that facial expressions say less about what people actually feel and more about how we seek to influence other people, say, by putting on a fake cry to gain sympathy (Crivelli and Fridlund, 2018). Not that it always works, but facial expressions and bodily gestures are often skilfully managed in order to disguise rather than disclose inner thoughts and feelings. At other times, we simply find ourselves unconsciously imitating the persons we are interacting with (e.g. Dimberg et al., 2000; Wood et al., 2016).

Regardless of what the underlying causes of particular facial expressions and gestures are, we suspect that we will learn more about the people we study if we appreciate that their interests, attitudes, perceptions and experiences may be expressed through their bodies as well as through the words they utter in interviews and conversations. The challenge is how to write about this in research accounts. Rather than ticking off boxes on coding schemes or insisting that a certain facial expression or gesture has a specific and determined meaning, we would go for the following strategy: to try and describe these expressions in as much detail as it makes sense to do, whilst leaving sufficient openness so that readers can develop their own interpretations. This is certainly a strategy that is routinely, and effectively, pursued by writers of literary fiction. Charles Dickens for one was prolific in his use of facial expressions, a typical example being 'How dull you are tonight!' sneered Miss Squeers (in *Nicholas Nickleby*, 1838, p. 238). Apparently, Dickens used the expression 'sneered' 35 times in his works (Nishio, 2013).

We have no illusion that this strategy will exhaust the true meaning of embodied, non-verbal communication, but it may bring life to narratives and characters. Thus, it may help us write richer, more vivid and more engaging accounts. Instead of reproducing interview transcripts in the simple manner of question-and-response or in the convention of adding interview responses to back up analytical claims, themes and sub-themes, why not try and write up interviews and conversations as the social, face-to-face and embodied encounters that they are? Let us offer an example of what this might look like.

A face-to-face interview with a recovering star performer

Shortly after Torkild started his research project on corporate performance cultures he interviewed someone with substantial experience of working in this kind of environment:

It's a grey Wednesday morning. We're in the middle of October. My wrist watch shows 10am but it's really 9.55. We've agreed to meet in a cafe in one of the business districts on the outskirts of town. I'm waiting by the entrance. I don't know much about his background, only that he's working for one of the up-and-coming consultancy firms. I also don't know what he looks like, so I'm a bit surprised when a guy in a baseball cap makes eye contact as he walks slowly towards me. He shines up when I stutter his name and stretches out his hand in a firm but gentle handshake.

'Just a black coffee for me please', he replies as I head to the counter.

He strikes me as calm and composed. Confident but not arrogant, despite the posh accent. He's quite tall, probably two or three inches taller than me, and just as generously built around the waist. It suits him. The beard suits him too. It makes him look mature for his age. I suppose I'd expected a lean and clean-shaven 35-year-old in a slim-fit business suit. I'm right about his age, but his attire is casual. Elegant but casual: designer sports coat, designer trainers and, from what I can see, designer jeans.

'This is actually my day off', he says, taking a sip from his coffee. 'I work Monday mornings, Tuesday mornings and Thursday mornings. Then I have a day off both Wednesdays and Fridays.' He leans back in the sofa.

'I thought your firm had a high-intensity performance culture', I blurt out. 'That you worked all the time?'

'That's right, but ...'. He smiles. My indiscretion doesn't seem to have bothered him. 'I was burnt out a couple of years ago. So, I'm working twenty-five per cent at the moment ... I might be able to go up to fifty per cent later on.'

It finally dawns on me. I thought he looked a bit tired.

'But before that I very much lived that performance culture. I worked sixty, seventy, eighty hours a week.' He leans forward, his voice drops: 'Until I collapsed, and woke up with blood on my shirt. I was unable to read for two months. I spent several weeks in hospital.'

'But ... blood on your shirt?'

'Yes. I fainted and fell, hitting my head on the banister. It was a stress reaction. I had all the warning symptoms for burnout syndrome ... all twenty-one of them.'

'So, this was both a physical and mental reaction?'

'Yes. I didn't see it myself, or, I didn't want to, though people around me did notice what was happening to me ... Colleagues, family, friends ... They told me that perhaps I didn't need to work as much as I did. That, perhaps I'd like to do other things besides work. Take time off once in a while, spend time with family, have fun ...'. He pauses for a second or two. 'I told them that maybe they ought to work more!'

(Continued)

(Continued)

He looks back at me, as if to make sure that I get it. I hadn't expected this. I'm perplexed by his story, and I'm impressed by how articulate he is, particularly as I'm starting to realize what he's been through.

'This was my wake-up call. That I can't work like I used to. This has given me a different perspective on life …'. He pauses. 'I had a rather naïve attitude to myself. I thought everything was fine just because I kept pushing myself … I could work 10 hours non-stop, without a single break. I'd eat something of course, and go to the bathroom when I had to … my eyes would get completely red. I wouldn't be able to see properly, but I still kept on going.' He takes another sip from his coffee. 'If someone told me that something was impossible, that was the worst thing they could say to me, because then I'd be determined to prove them wrong …'. He seems more energetic now, and speaks passionately about his condition. 'Being a star performer became a drug. Now that I've read about alcoholism I've realized that I was addicted to top performance. To deliver it I mean. The challenge became my bottle ... So, I need rules, just like a recovering alcoholic ... You know, New Year's Eve is hard work for an alcoholic. For me, it gets rough when I'm around challenges. I still notice that I can get a kick out of performing.' He slows down. 'But I tire quickly too. I can't do the same jobs that I did. I only do simple tasks, but even that can go wrong …'

'How?'

'For one thing, I can't keep track of numbers like I did.'

I wonder what to say next, and I think he notices, because he suddenly interjects:

'I can tell you as much as you like to know you know. You can ask any question you like.' His candidness baffles me. 'But you can't mention the name of the company. It can't come out that I've been talking to you. They're quite secretive you know. And if you mention the name of the company it'll be fairly easy to find out who I am.'

I promise him that I won't, that my study has taken a different turn since we were in touch before the summer, that I'll primarily use the interviews to understand the bigger picture of corporate performance cultures. He nods, and seems pleased with my response.

Our conversation turns to his background, and to the company. He tells me about their business model, their competitors, the organizational structure, the pay structure, staff turnover and the culture. He tells me about the extreme tempo, the young ambitious staff and how anxious they are – about fitting in, about performing at the top level no matter what, about always looking their best.

'It has always bothered me that they party so much', he says, frowning. '… and their fixation with looks … but what's the point of looking good if you throw up your lunch?!' He lights up: 'You know, some of them barely eat!' He gets self-conscious, and stares at me while lowering his voice: 'They snort coke instead … to be able to keep up with the pressure.'

We had agreed to go on for an hour, but at this point we've talked for an hour and a half. He seems invigorated by our conversation, so I ask if I may take him to lunch. As we're joining the queue, he looks at me:

'I'm sorry, Torkild, but I don't think I'll be able to do lunch anyway … It's time for me to get back. You see, I usually lie down for an hour around lunchtime.'

I've been too focused on our conversation to realize that he actually does look a bit worn out from our meeting. We say goodbye, and I promise to keep in touch.

Describing and revealing the body

Naturally, there are risks and limitations to this method. Although Torkild did his best to include many of his interviewee's gestures and facial expressions like they happened when they happened, for stylistic and cognitive reasons he took certain liberties with other details. His interviewee did greet him with a smile and with a firm yet gentle handshake, but he did not necessarily sip his coffee or lean back on the sofa exactly at these points in the interview which Torkild's account seems to suggest. It may not be aesthetically appropriate or indeed possible to recall and reproduce every gesture, movement or facial expression with perfect accuracy. That is also not the point. Taking inspiration from literature means that there will always be an element of fiction and stylistic manipulation. As the late science-fiction author Ursula K. Le Guin (2017) reminds us, fiction is not the opposite of truth. Fiction does not lie. Whereas lies claim to be facts, fiction does not.

Furthermore, this method raises ethical challenges about how we may handle confidential information and protect the anonymity of our interviewees. If the portraits we make of people reveal details about their looks and body language, will not that make it easier for readers to identify who they are? The stakes get particularly high if we include sensitive information, like Torkild just did, and he may have made things worse by exploiting someone who was in a vulnerable position. We like to think that he didn't. Despite his burnout, Torkild's interviewee struck him as resourceful. Articulate and confident, reflexive and analytical. While he stressed the need to remain anonymous, there was little reason to doubt that he wanted to share his experience. But as other things got in the way, Torkild waited almost a year before he emailed him a copy of this narrative, to check that he was okay with it, an issue we will say more about in the next chapter when discussing the ethical and practical concerns of recording looks, facial expressions and body language while interviewing others.

This method yields opportunities too. If social scientists should aim to have a positive impact on the people and living conditions that we study, we doubt it will hurt to share stories that highlight the gravity of social and organizational problems, such as stress and burnout. Research accounts that affect and move readers may be more likely to inspire change and action. Personally, we are certainly more strongly moved if we get a sense of who the story is about and if the narrative makes us feel that we know the people we read about, if only indirectly, and just a little bit.

To do so, many of us may need to change how we write. The quasi-quantitative tendency of reproducing a large number of short quotes from a large number of respondents needs to be challenged, perhaps particularly because this has become such a widespread strategy for publishing empirical research in academic journals. Without ignoring the increasing pressure to publish in highly ranked journals which especially confronts early-career academics striving to get permanent positions with tolerable teaching loads (Clarke and Knights, 2018), academic journals are neither the sole publishing outlet for academic research, nor are their conventions of writing fixed, immutable and incontestable. We shall return to this issue in Chapter 8, when discussing the challenges of writing, disseminating and publishing embodied research. But before we move on, we would like to summarize the key points and arguments that we have put forward in this chapter.

SUMMARY: INTERVIEWS AND CONVERSATIONS ARE EMBODIED ENCOUNTERS

The embodied approach that we are advocating here accepts many of the basic presumptions and techniques employed by those following more conventional approaches to the qualitative research interview. Since interviewees may be pressed for time it is advisable to reconfirm the agreed length and format of the interview. However, the first challenge is to seek to create an atmosphere where both are feeling relaxed. This can be difficult as often a more senior person has granted the access and the interviewee feels simply a hierarchical obligation to agree to the interview, regardless of the researcher stressing its voluntary nature. Secondly, it is a good idea to remind interviewees about the purpose of the research and promise feedback so that they can feel to have made some contribution to knowledge. Unless told explicitly otherwise (as reported by David in the previous chapter), interviewees often want to know what will happen to the interview material. At the same time, they may be concerned about revealing sensitive information, so they should always have the opportunity to remain anonymous.

Furthermore, we fully recognize that verbal content is generated by asking and answering questions. While this makes it sensible to consider what questions to ask and how to ask them, it is just as important to try and be empathetic so as to encourage open and embodied reflections from interviewees as they tell you their stories. It is unlikely that you will be able to ask all of your questions or cover all the issues of your concerns, but often the best interviews are those in which you ask hardly any questions other than those that prompt further elaboration of what they are already saying.

Acknowledging how difficult it is to manage any social encounter, these matters are likely to affect our ability to build trust and rapport with our interviewees, initiate conversation and get people to talk. Seeking to impose too much structure and order risks closing down the conversation and, of course, asking questions can do precisely this, so the objective should be to have free-flowing conversation, although necessarily a bit one-sided. Since people do generally like talking about themselves as long as it carries few risks, the aim is to convey the encounter as non-threatening as possible.

But more importantly, our discussion in this chapter suggests that interviews and conversations are affective and embodied encounters between people, which trigger emotional responses, actualize power relations, and where much of what is communicated is conveyed through our gestures, facial expressions and body language. Much more happens in interviews and conversations than asking questions, answering questions and listening for verbal content.

Questions and exercises

1. In what ways may interviews and conversations constitute sources of non-verbal communication?
2. What advantages and disadvantages do you associate with taking an embodied approach to interviewing?
3. Interview a friend, colleague or fellow student for 15 minutes. Agree on a topic and try to keep notes during the interview. How did it pan out? What would you have done differently? Once you have evaluated this interview process, reverse the exercise to learn what it might be like to be interviewed.
4. Team up with two friends, colleagues or fellow students. Agree on a topic and a division of labour so that one person acts as the interviewee, one person acts as the interviewer, and the third person focuses on observing the interviewee's facial expressions, physical gestures and body language. Now, let the interviewer and the observer compare notes and write up the conversation in a way which even portrays some of the interviewee's facial expressions, physical gestures and body language. Rotate the roles and repeat the exercise twice so that each person gets a chance to practise each role.

6 RECORDING AND DOCUMENTING EMBODIED LIVES

As we saw in the previous two chapters, there are a number of challenges associated with recording and managing empirical material generated through research interviews and ethnographic fieldwork. For instance, how do we record our observations if we are not able to take photographs or make video-recordings? What if participating in the activities that we are meant to observe makes it inconvenient or even impossible to use a notepad? What about conversations and interviews? How do we keep note of what is being said if we are not allowed to or able to audio-record it? Those of us who are keen to do embodied research might find these things particularly challenging; we are not only interested in what people do and say but also how people look and act with their bodies.

While recording and documenting your research is crucial in order to know what material you have generated, it is equally an important aspect of managing relations with research participants beyond field visits and interviews. In our experience, it is difficult to report back to interviewees and research participants unless you have decent records from your meetings.

In this chapter, we shall discuss some of the methods that may be utilized to record and document embodied lives, including photography and video, note-keeping, audio-recording and transcribing. Further on, we will touch upon the role of memory and discuss how biographical methods such as diary-keeping may help us and our research participants remember.

DEPICTING BODIES THROUGH VISUAL MEDIA

Over the past decade or so, visual methods such as video and photography have become increasingly popular among social scientists. Although much of this research focuses on analysing photographs, videos and other imagery as visual materials (see, e.g., Rose, 2016), video and photography may even offer effective means of documenting the bodily movements, physical gestures, appearance and feelings of research participants. This has certainly become easier and cheaper as most of us have simple camera and video-recording technology available through our smartphones.

Anthropology has a long albeit problematic tradition of using film and photography to record and document the bodily appearance, rituals and daily activities of people and cultural communities. In early anthropology, photographs did not merely make it possible to display features and habits of geographically distant and isolated cultures to a western audience which unlikely had come into contact with these cultural groups themselves. The proto-anthropological research carried out in the latter half of the 19th century was deeply embedded in the time's colonialist and racist zeitgeist, and scientists such as Thomas Huxley and John Lamprey used photography as an 'anthropometric' method in order to 'document supposed [morphological] differences between racial groups' (Emmison et al., 2012, p. 29).

By the 1930s, anthropology had become an established social science discipline, and leading anthropologists such as Margaret Mead and Gregory Bateson used photography and film to record bodily activities and body styles in non-western cultural communities. In his previous attempt to study the Baining of New Guinea, Bateson had found them secretive, and he had struggled to gain sufficient access (Lipset, 1980). In Mead and Bateson's study of Balinese culture in the village of Bajoeng Gede, Bateson had been able to take 25,000 photographs while Mead had kept thousands of pages of field notes (ibid.; Silverman, 2005).

This massive number of photographs was not generated without practical difficulty or mischievousness. By today's standards it may be argued that Bateson violated basic principles of research ethics through his sometimes invasive and unscrupulous use of the camera, using an angular view finder on his stills camera to frame shots of people while they were eating and at other times when they would turn their backs to avoid being photographed (Bateson and Mead, 1942, p. 49; see also Banks, 2001, pp. 120–1). As Bateson and Mead admit in the caption to a photograph showing a group of Balinese men eating; 'The eating of meals is accompanied by considerable shame. Those who are eating usually turn their backs toward anybody who may be present' (Bateson and Mead, 1942, p. 112, in Banks, 2001, p. 121).[1]

This does not mean that the use of film, video and photography in social science research needs to be rejected as ethically dubious. Recording and documenting people's bodies, behaviours and social interaction through visual media do raise a number of other ethical challenges, but with the rise of these methods social scientists have also become more conscious and conscientious about how to deal with them. Before returning to these issues, let us look more closely at three examples which further illustrate how visual media may be used in embodied research.

Filming bodies in urban space

William H. Whyte's (1979, 1980) documentary film *The Social Life of Small Urban Spaces* offers an interesting case in this respect. In order to find out what attracted people to use certain spaces rather than others, Whyte and his colleagues set up time-lapse cameras in 14 locations across Manhattan, including the Seagram Plaza on Park Avenue, the entrance to the New York Public Library on Fifth Avenue and a block on 101st Street in East Harlem. Whyte's film mostly contains long shots and very few close-ups. Complementing recordings from the time-lapse cameras with direct observation and careful mapping techniques, the research team was able to identify patterns in how people used these places.

Hypothesizing first that people would be more likely to use sunlit places, this later turned out to have little effect, and they eventually concluded that people were more likely to use squares and plazas where they could sit whilst at the same time being close to the action of pedestrian street traffic. According to Whyte, chairs and benches are not necessary – steps and ledges of an appropriate dimension will do just fine – but it does help if it is possible to eat and talk in the plaza, if there are trees around which provide shade and seclusion, and if the plaza contains water, which is calming to the ear, pleasing to look at, and which affords people an opportunity to cool off their hot, swollen feet.

Whyte and his colleagues, then, set out to investigate the relationship between spatial design and social interaction. However, the film also offers a powerful illustration of how people inhabit urban space through basic bodily activities of walking and sitting, eating, talking and watching. It shows how the social aspects of urban life are entangled with our bodily needs and preferences, in sometimes-unexpected ways. We highly recommend watching the film. When we were writing this book, it was easily accessible online.

Photographing feelings in office culture

Our colleague Samantha Warren's (2002) study of office culture offers somewhat different insights into the use of photography in embodied research. Whereas Whyte's film generates a largely distanced view into the bodily movements and activities of the people who find themselves involuntarily participating in his film, Warren's (2002) use of auto-driven photography and photo elicitation (Collier, 1957; Collier and Collier, 1986) seeks to visualize embodied feelings without depicting any human bodies.

As part of a broader study of office culture and office aesthetics in a UK IT company, Warren provided employees with a digital camera and asked them to 'take a set of photos ... that the respondents felt represented their work environment to them' (2002, p. 232). She then met with her participants and asked them to 'Show me how it feels to work here' (p. 232). This generated a number of photos – of cubicles and corridors, of chairs and sofas, of cookie bags and door locks, of trees, plants and the sky outside the office building, and of enormous inflatable 'boss' figures made in the style and shape of Russian dolls. While no people appear in most of the photographs published in Warren's account (there are certainly no close-ups of people), an emotional expressivity is produced by the combination of image and text. When Warren met with her participants they told her how they associated the Russian dolls and door locks with feelings of control and oppression, but also that they had photographed cubicles and cookie bags to illustrate a sense of community spirit. As they commented on the photos, they spoke of their colleagues with pride, pleasure and love.

Photographing bodies in a corporate performance culture

In Torkild's study of corporate performance culture he utilized photography in a much more basic way:

> Out of concern that it would interfere too much with my research participants' work, I wasn't able to actively involve them in generating research material or engage them in any genuine conversation about the photos I've taken of them and their work environment. In order to record and document what it may be like to live and work in a corporate performance culture I've primarily taken long shots and close-ups of employees during more or less extraordinary events. Thus, I've got several photos from beach volleyball tournaments, eating contests and fitness tests, but also from morning meetings, sales meetings and staff training sessions. Although these events may not give a representative image of an average work day, I still think they express something about the organization's culture and what it's like to work there.

Torkild's concerns find some resonance in Becker's (2002) view that photographs and photographic sequences in film 'embody ... specified generalizations' (p. 5). According to Becker, it is not possible to photograph a general phenomenon such as contemporary capitalism, but it is possible to photograph 'people and things who ... exemplify or symbolize or embody these ideas' (p. 5). Similarly, it may not be possible to perfectly represent or capture all elements of the performance culture Torkild has studied in one photograph or photographic sequence. However, just one photograph of a team or a team member may illustrate and embody the wider phenomenon of corporate performance cultures. It could even be argued that it illustrates how the general idea of contemporary capitalism is played out in people's everyday lives.

In contrast to Warren's (2002) published images (where employee feelings were expressed through photos of objects, artefacts and office interiors), people figure prominently in Torkild's still unpublished images:

> My photos show people's bodies and faces – as they're giving or attending a presentation, as they're slamming (or missing) the volleyball, as they're stuffing their face with cream buns (or cheering at a colleague stuffing theirs), and as they're trying to beat company fitness records by doing as many burpees and push-ups they can in two minutes. In some of the photos, employees strike a pose, but in most of them they don't. Since I'm quite a terrible photographer, many of my photos are out of focus, poorly lit and taken at awkward angles. Thankfully, it hasn't been my aim to take high-quality photographs. Rather, I've strived to record lived, felt and embodied experiences of working life by getting snapshots of people as they're involved in naturally occurring events and activities – of their facial expressions, gestures and postures, of bodily activities and face-to-face encounters.

This inevitably raises ethical concerns which may be avoided by following Warren's (2002) strategy. In cases where researchers have promised to honour participants' anonymity and integrity, it would be risky, even foolish, to publish or publicly share photos that make participants identifiable. This is a pressing issue in the kind of research that Torkild has been doing, as many of the photos portray people and their bodies in intimate and sensitive situations, when they might find themselves vulnerable: as they are running out of breath during a fitness test; as they are failing to improve their fitness performance from last year; or as their face is covered in cream, only just having realized that they have lost the eating contest to a co-worker.

The publication of most kinds of research materials requires some kind of explicit consent from the participants themselves, but gaining explicit and unequivocal consent from participants is essential when considering if one should publish photographs that make participants identifiable. While Torkild has shared the photos he took with the individuals appearing in them, most visual scholars would regard this a minimum of researcher responsibility, which does not automatically grant the right to publish them. However, photographs taken of research participants do not need to be displayed and published in order to be of use in embodied research. This does not relieve any of us from the need to gain permission from the individuals we would like to photograph, but it does make it much easier to protect participants' anonymity. Using photography to record the embodied activities and interactions of research participants may help us keep track of our observations.

Moreover, having one or several photographs from an event or an activity might enable us to develop a richer understanding of that same event or activity. Looking carefully at a photograph you may be able to spot things which were not immediately apparent when you took the photograph or when you attended the event depicted in it. So, for example, when studying vets, David and his colleague Caroline Clarke used photographs which reminded them of the bodily challenge that handling large animals such as cows weighing around 600 kilograms might present. It made more vivid statements from vets about shifting to small animal vetting because of the danger of injuries to the body. But Caroline also took photographs to record surgical procedures, such as the two images shown in Photos 6.1a and 6.1b, which show two vets operating on the hernia of a calf: the first photograph shows how one vet stands by while the other shaves a part of the calf's belly before operating; the second photograph is a snapshot from the actual operation.

(a)

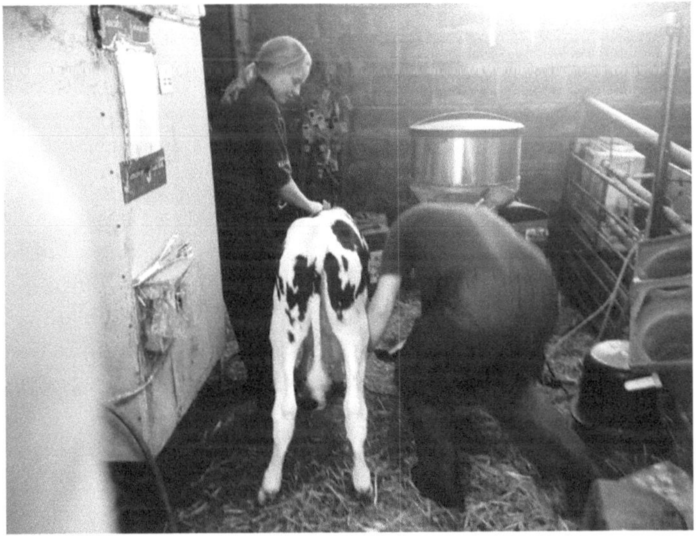

(b)

Photo 6.1 Photographs by Caroline Clarke.

In a pioneering commentary on the purpose and practice of visual sociology, Becker (1974, p. 7) shares an exercise that he was taught by the professional photographer and photography professor Philip Perkis. It goes as follows: take a photo and look at it actively for two minutes (don't just stare); name everything you see and continue to do so for five minutes; then 'tell yourself a story about the people and things you see in the picture'; now ask yourself what 'emotion and mood the picture has evoked'. As you do this exercise several times, Becker argues, your 'careful way of looking becomes habituated', and you will 'develop new habits of seeing'.

Torkild tried to do this exercise with a photograph he had taken during his research on corporate performance cultures.

An exercise in careful looking

I took the photo on the morning of 4 April 2017. I'd come to the downtown gym where the sportswear company do their weekly sports hour, but also their biannual fitness tests. This morning the CEO was up for his. He had invited me to come along the previous day. I remember asking him if it was okay that I shoot some photos of him: 'Take as many as you like!'

When I look through the 314 photos on my computer nearly six months later, I notice how out-of-focus many of them are. Rather than selecting one at random for closer scrutiny, I try to pick one that I hope I'll be able to say something writable about. Perhaps I've already ruined Perkis's exercise. Anyway, I finally settle on a photo taken when the CEO was halfway through his fitness test:

> We're in the downstairs workout area, in front of the bright pink entrance to what I deduce must be the women's changing rooms (I remember that the entrance to the men's changing rooms is painted turquoise). With the exception of one patch of wall, which is white, the rest of the walls are nearly black (charcoal?). So is the hard rubber floor. Some kind of decline bench or back extension bench is pushed against one of the walls. A weight bar is left carelessly on the floor. Two rails, or ladders, are fixed horizontally to the ceiling. There are four people in the picture: two women (who are sitting on the floor minding their own business while stretching out), one male gym staff member (who is leaving the room with a bundle of exercise ropes) and the CEO.
>
> The CEO is in the centre of the picture. He is wearing a pair of navy-blue running tights and a blue t-shirt. The t-shirt is creased by his pectoral muscles, and it is patched with sweat from the previous exercises – on his stomach, under his pecs and around his armpits. His left arm is hanging down in front of his left thigh. He is bending his right arm so his right hand covers part of his waist. This is his bad arm. He pulled off his bicep skiing off-piste in the Alps, five or six weeks earlier. There are two ropes in front of him, dangling from the ceiling. But he is looking up at the horizontal rails mounted above his head. Mouth half open, he is anticipating what awaits him: one-arm pull-ups. Had his right arm been all right, this would have been a piece of cake. With the use of two arms he'd do 25 pull-ups, easily.

I should admit that I didn't follow Perkis's exercise slavishly. I didn't time the exercise, and I didn't start off by listing all the things I saw in the photograph. Instead, I looked at the photograph, and started writing my story of what I almost immediately saw after I began looking at it and scrutinizing it. Rather than just writing, I had a vague idea of what I wanted to say. This probably ended up taking me close to an hour.

Could I have said these things without the photograph at hand, based only on my notes? Possibly. But I did notice some things that I doubt I would've been able to recall had I not had a photographic record if it. I had no idea of how many other people were in the photograph, and I would've had to go back to the venue if I had wanted to describe its spatial features in any detail. Regardless, without the photograph I don't think I would've remembered the CEO's bodily posture or his exact facial expression at this moment in time, how he was positioned in the room or what and who was behind him and around him.

Perhaps this wouldn't have mattered much if I'd simply wanted to offer a brief description of the company's fitness tests. From my notes I could've reproduced parts of the conversation between the CEO and the personal trainer, and I could easily have reproduced facts about the various exercises he did and how well he did at each one. I could even have shared basic observations about his excitement at the beginning of each exercise, and his exhaustion as he was completing them. However, it would've been difficult for me to describe each scene in detail or portray his embodied presence with much nuance or intensity.

When looking at another photo from the CEO's fitness test, I'm reminded of the effort he puts into each exercise. In order to measure his flexibility, the personal trainer has instructed him to sit down on the floor to see how far he can reach beyond his toes whilst keeping his legs straightened. She has placed a plastic step module over his legs, to get a reliable measurement, and to prevent him from bending his knees. His face is wrenched in pain, as he struggles to gain an extra inch on the measuring tape. I look up the photo one more time on my computer and notice that his fingertips are barely reaching his toes.

TRUTHS, LIES AND RICH DETAIL

To us, these examples suggest that photography can help social scientists craft richer narratives from our research, even if we do not choose to publish our photographs as part of our books and articles. However, does that mean that photography makes our research accounts more truthful? That you don't have to take my word for it because there is visual evidence too, which verifies my truth claims?

Now, if the truthful representation of reality was the *sine qua non* of photography in embodied research, why stop at photography, which only utilizes one sense? Surely, a complete and absolute representation of the embodied social world would require us to generate empirical material using all our senses – not just vision and hearing, but even smell, touch and taste.

In a remarkable study of office life, our colleagues Kat Riach and Sam Warren (2015) developed an olfactory method to investigate how employees deal with their experiences of smell at work. Utilizing videotaped smell interviews, paper strips and audio smell diaries, they were able to engage research participants in vivid conversations about the various smells (and the absence of smell) which they associated with work, with specific colleagues and with the workplace – smelly sandwiches, strong perfumes and body odours. While this olfactory method produced an enticing narrative about workplace smells, it was not free of the technological constraints of academic publishing. The odours and smells that were talked about were only described and neither emitted nor smelt in the published article, and as readers we were left to imagine these smells rather than actually smell them.

There has been a long debate in the history of photography about these epistemological issues, spanning from Henri Cartier-Bresson's position that 'Our task is to perceive reality' without manipulating it (in Goldstein, 2007, p. 62), to Lewis Hine's admission that 'while photographs may not lie, liars may photograph' (p. 62). Closer to Hine than to Cartier-Bresson, the photographer Barry Goldstein offers a rather blunt claim, which nevertheless summarizes our most pressing concern here: that 'All photos lie.' 'Every photograph lies' (2007, p. 64), he maintains, because 'Cameras cannot replicate human vision' (p. 65); because 'A photograph only records a brief moment of time' (p. 70); and because 'Every photograph is manipulated' (p. 75) – by the various choices the photographer makes in deciding what to photograph, why to photograph it, and when and how to do so using what equipment.

Does this mean that there is no value in utilizing photography in embodied research? No, but this is a useful reminder that photography, like any other visual or non-visual research tool, can do certain things for us, but not others. Its representational limits should not discourage us from experimenting with its expressive potential or exploiting its capacity to record, store and help us recollect empirical incidents and encounters. It is the expressive capacities of photography rather than its representational constraints which give us hope that the stories we tell of social life, and of our lived experience thereof, can be made livelier, more interesting and more worthwhile reading.

As we have hinted above, photography offers other advantages too to the embodied researcher. Through the technique of 'auto-driving', whereby research participants are encouraged to take their own photographs, you may be able to generate more images, and a wider variety of images from a wider variety of settings, especially from settings that may be off limits to you as a researcher (see, e.g., Young and Barrett, 2001). And by involving participants in photo elicitation (e.g. Collier, 1957; Collier and Collier, 1986) you may be able to get comments on the visual material without which it may be difficult to make sense of it (e.g. Young and Barrett, 2001; Warren, 2002; MacDougal, 2005; Ray and Smith, 2012; Greenland and Rayman-Bacchus, 2014; Pink, 2015). Furthermore, auto-driven photo elicitation may prompt rich conversations with initially reserved research participants (e.g. Slutskaya et al., 2012). Finally, it has been argued that photographs may help researchers feel empathy for research participants, at least in fields where researchers and practitioners are under tremendous time pressure, or in disciplines where students are traditionally trained to remain distanced and cool (Lorenz, 2011).

Despite their challenges, then, photography and other visual methods offer a number of advantages for embodied research in terms of generating and storing empirical material. However, what to do if we find ourselves in research situations where we are prevented from making visual records, when it may be ethically inappropriate to do so, or when it is not even practical to do so?

REMEMBERING, FORGETTING AND TRANSCRIBING EMBODIED ENCOUNTERS

As we have indicated above, researchers' easy access to visual and other recording technologies gives no automatic licence to use such technologies in research situations. While video-recording and audio-recording devices which can be operated hands free may help researchers get more intensely involved as interviewers, conversation partners or active participants, it is not always possible or appropriate to use them. Participants may visibly seem uncomfortable with the prospect of having their photograph taken or their voice recorded, or they may explicitly refuse any visual or audio-recordings from your encounter. In such situations, researchers may feel pressed to keep as full notes as possible during the research activity, fearing that failure to do so will leave them with no usable records whatsoever.

However, usable records are not the same as complete records, and incomplete records can still be of use. A risk with the most recent innovations in visual and audio-recording technology is that we mistake our increasingly high-quality recordings with the reality we are trying to capture with them. Of course, no recording is complete, and no representation of reality can fully capture or correspond to the piece of reality that we have photographed or audiotaped. There will always be a gap or slippage between the two-dimensional photographic image and the visual perception which preceded it, between the audio file and the conversation it recorded.

Furthermore, we should not forget that most research situations are social encounters. While it is obvious that Torkild could not possibly keep notes whilst participating in the sportswear company's fitness activities, it is perhaps less obvious why we might refrain from keeping notes during an interview where we have been prevented from audio-recording the conversation. But, trying too hard to get full detailed notes during an interview risks ruining the conversation, giving you little material to work with in the first place. For the same reason, we would not be too vigilant in trying to annotate facial expressions and gestures during an interview where we had the use of an audio-recording device. Rather, we would rely on our fallible memories. This was exactly what Torkild was faced with when interviewing the recovering star performer, an account of which we reproduced in Chapter 5:

> Since he was revealing sensitive information about the company's extreme performance culture and about his own burnout syndrome, he was determined to stay anonymous. What he told me, he told me in confidence. And since I was desperate to know as much as possible about him and the company, I didn't even dare to ask him permission to audio-record our conversation, out of fear that he might pull out of the interview altogether. I also didn't take full notes during our conversation as I sensed that this might risk interrupting the flow of our conversation. Instead, I ended up sitting down with my notebook immediately after I came home. While I worried that I would remember very little from our conversation, once I'd sat down on the couch in my study, my head filled with his words, and with mine. Mixed with flashing images from our meeting, one recollection triggered the next, not in perfect linearity, but close enough. Two hours later I had 16 full pages of handwritten notes, reproduced from memory.

David had to resort to a rather different approach when doing fieldwork in the newspaper industry for his PhD:

> I was doing observer participation among print union staff, which mostly involved situations of illegitimate social drinking episodes carried out during working hours. Since I felt it not appropriate to audio-record the events *in situ* I could only write up copious notes later or by frequent visits to the toilet. As often I would still be under the influence of alcohol from participating in their drinking sessions, my notes were somewhat distorted. Nevertheless, I was able to record important issues surrounding what I described as three levels of workplace deviance, or what the staff called 'skives', 'super skives' and 'super-duper skives'. Insofar as I was able to observe and recollect not just their actions but also their bodies, it was possible to see how, in the name of excessive pleasure and recreation, these staff were suffering deep discomfort if not pain and sadness in violating the routine norms and standards of everyday working life and employment.[2]

We acknowledge that our memories might fail us, for it often plays tricks on us, so we do not insist that others should copy this 'method'. But if the situation prevents you from audio-recording interviews and simultaneously keeping notes of bodily features and body language, we believe this is an okay alternative, which is not much different from established practices of note-keeping in ethnographic fieldwork. Indeed, one reason why interviewing and participant observation have been thought to require different recording practices may be that they have been seen to (mistakenly in our view) generate different kinds of material. This is a flawed view, which ignores that even interviews are social and embodied encounters which involve non-verbal as well as verbal communication.

So far in this chapter we have shown how embodied practices and experiences can be recorded through a variety of different formats, from visual technology to note-keeping. Let us now move on to consider how biographical forms of writing may help us delve deeply into the most intimate, private and sensitive realms of our lives. The diary and the memoir are two such genres, which can help researchers record our own embodied lives, or be used by research participants to record theirs.

EMBODIED INTIMACY IN BIOGRAPHICAL ACCOUNTS

Biographical accounts such as diaries and memoirs are obviously verbal forms of expression, yet they may enable research participants to write about events and experiences from a more intimate and embodied perspective. This may be particularly true of the personal diary.[3] While participants may be unable or reluctant to share and reflect on personal experiences and felt emotions in interviews, they may be more prepared to do so in a diary. According to the feminist historian Elizabeth Hampsten (1989), diaries may facilitate such narratives because they tend to be less tainted by the assumed status and power of the researcher. Moreover, diaries are neither constrained by the short time frame of the interview nor dependent on the researcher's presence. Thus, they enable participants to record events and reflect about embodied encounters and experiences shortly after they happen. As some commentators have pointed out, diaries record life 'in an "immediate perspective"' (Watson and Watson-Franke, 1985, p. 3), making it possible for researchers to 'captur[e] life as it is lived' (Bolger et al., 2003, in Lewthwaite and Bartlett, 2017, p. 2).

At the same time, diaries that are kept and collected for research purposes are often tightly structured, limited in time span and focused around a specific problem or activity. For example, in a recent study on people living with chronic back pain, Broom et al. (2015) asked their research participants to keep a structured diary over the course of one month. Participants were requested to fill in their name and the date they started their diary, and they were instructed to complete a number of specific questions every day: about medication, health checks, daily activities and feelings of pain and wellbeing. While this generated mention of mundane activities such as cleaning the car and going for beach walks, participants also gave brief yet direct descriptions of feelings of pain and fatigue, of sleepless nights, and of the frustration and hopelessness experienced from not being able to live life and do all the things they would like to do.[4]

Although structured research diaries may offer important insights into the lived experiences of research participants, the capacity of diaries to yield an immediate perspective into the writer's life may be better illustrated by examples outside of academic research. Samuel Pepys, who was a notable English civil servant and naval administrator in the 17th century, is regarded to have kept the first modern diary. Between 1660 and 1669 he kept regular records of his daily routines, activities and concerns, including business meetings, dinners and love affairs. Without endorsing the lifestyle of this philandering member of the English elite, it is worth noting how Pepys writes in a matter-of-fact style which nevertheless expresses his joys and pains, his qualms and fears, with vivacity and some self-irony. For instance, on February 8th, 1661, he writes how a full morning at the office followed by noon meetings at the Exchange ended with the evening spent with his old friend Dick Scobell at Mr Rowlinson's, where 'I drank a great deal' before getting 'home and to bed betimes, my head akeing' (Pepys, 2003, p. 118). In late October and early November 1668, records of his business matters are entangled with worrying and self-pity, as he realizes that he will need to fire his mistress Deb Willet from the household now that his wife has found out about their affair (pp. 950–60).[5]

The form and content of diary-writing remains contested, and other prominent diary writers, Virginia Woolf to name one, have problematized what a diary should be and what it should be about. In the diaries that Woolf (1990) kept between 1915 and her death in 1941, she is quite economical with how much she shares in terms of embodied experience and intimate reflection. At least on the surface, details about her emotional life and inner feelings give way to routine yet elegant notations about daily activities (work, walks and meals), current affairs (worries during the two World Wars and anticipation of peace), the weather (mostly rain), and toxic but humorous portraits of friends and family (especially her mother-in-law and her close friend John Maynard Keynes).[6]

Less than three weeks before Woolf drowns herself in the River Ouse near her Sussex home she insists that 'I intend no introspection: I mark Henry James's sentence: Observe perpetually' (8 March 1941, p. 503). However, Woolf admits that her observations are not entirely of the exterior. They 'verge ... on introspection' and are turned towards her 'own despondency'. Yet, her despondency 'becomes serviceable', she notes, insofar as she pens it and keeps occupied. Writing her daily routine and 'letting things come one after another' prevents her from disappearing into a depressed abyss of introspective self-pity and self-indulgence. Thus, she notes 'with some pleasure' that it is 7pm, and that she 'must cook dinner. Haddock and sausage meat.'

Indulging in ourselves, and adding to others' pain?

Woolf's concerns about introspection are perhaps one reason why diaries are infrequently used by social scientists and other scholars, even those of us eager to generate explicitly embodied accounts of social life. When we discussed this chapter over the phone, Torkild remembers asking David if he might want to keep a diary while recovering from his heart surgery:

> 'You know, this is sometimes used in medicine and psychology, to study how patients experience their medical condition, how they handle it ... Our readers may be interested to learn about this from your perspective don't you think, to get a sense of the physical pain you're experiencing ... how you're handling it?'

As Torkild was talking he noticed David's silence on the other end of the line. David seemed sceptical:

> 'It hasn't occurred to me actually. I suppose I've always felt that it would be too self-indulgent.'

Ignoring David's concern, Torkild backtracks yet pushes on:

> 'Well, I don't want to force you to do it if you don't want to. I mean, you shouldn't do it if you're too tired to do it, if you find it stressful. You shouldn't do it if it interferes with your recovery.'
>
> 'No, I'd be alright. I've already done some work ... We had to get those revisions done you see, on this paper for *Human Relations*. So perhaps I could try it.'

Looking back, we should admit that neither of us has ever kept a diary as such. However, Torkild's notes from his fieldwork in the sportswear business are intermingled with notes detailing everyday experiences whilst doing the research. Flicking through his first notepad from the project, the following entry made on 5 October 2016 grips his attention:

> I've got a cold and my neck is stiff. My knee is a bit better, but not well. Damnit. Will I be able to go to the sports hour on Friday? The doctoral course starts at 12 and I still have quite a bit to sort out. Updating the schedule and preparing my mini lecture on Merleau-Ponty. How will I manage? Difficult to lecture on something as complicated and abstract as M-P. How embodied was he really?

As we continued on this chapter, we did not seem to quite agree on the usefulness of diaries. On the one hand, Torkild kept asking himself why the risk of self-indulgence should stop us from keeping diaries:

> 'If we're interested in the embodied lives of other people, shouldn't we dare to venture into our own embodied lives too? Isn't there a risk otherwise that we pretend that only others have bodies but we don't? And may not the diary help me question myself, rather than affirm my self-identity?'

David's reply was toxic but sharp:

> 'I suspect that many diaries are written to do precisely the opposite. Aren't they a bit like selfies that are like a self-obsessed pandemic in contemporary society?'

However, there is no reason why research diaries should be limited to reflections centring on the researcher's own feelings and experiences, and even so, such reflections can be included economically in published material. Furthermore, research diaries may prove useful for recording body language and emotions among research participants, and to critically reflect on the research process (as, e.g., Burke did in her participant study of climbing Mount Everest [see Burke et al., 2008]).

Asking research participants to keep diaries generates other problems, which are rather apparent when for example reconsidering Broom et al.'s (2015) study of chronic back-pain patients, which we discussed above. Indeed, some of their participants had pointed out that they sometimes preferred *not* to think about their body, their chronic back pain and how they felt about it. Enlisting research participants to keep diaries, then, may risk adding to their pain, making the exercise taxing rather than therapeutic.

As these qualms illustrate, using diaries in embodied research has disadvantages and advantages. They certainly put pressure on researchers to cultivate our sensitivities – to be sensitive about what we ask research participants to share in a diary, to respect that some participants may not want to write everything in a diary, or to show everything they have written in it.

In *The Use of Personal Documents in Psychological Science*, Gordon Allport (1942) contrasted the diary from the memoir. Whereas the former is often not meant to be read by others, the latter is traditionally written for a public audience. Mind you, it is difficult to draw a clear boundary between diaries and other biographical genres. According to the military historian Yuval Harari (2008, 2009), certain memoirs may constitute forms of 'flesh-witnessing', that is, direct accounts of lived corporeal experience (see also Godfrey and Brewis, 2018). Harari (2009), a student of military battle, is actually quite critical of flesh-witnessing, viewing it as 'a rival authority' (p. 226) fundamentally different from scholarly accounts based on eye witnessing. Whereas the authority of a historical account of a military battle may be strengthened by the distanced accumulation of observed facts provided through eyewitness accounts, flesh-witnessing rests on the intense lived experience of someone who was actually there, directly involved in the battle yet unable to observe a vast amount of facts and often in disagreement with other flesh-witnesses.

Such blending of genres is evident in Juliet Jacques' (2015) recent book *Trans: A Memoir*, where she combines acute observations about gender politics in contemporary western society with candid, moving and beautifully written recollections and reflections about the ups and downs through her own transition. But it is even evident in what is widely regarded as the first modern autobiography: the *Confessions* of Saint Augustine.

At the same time, Pepys' and Woolf's diaries did not merely speak of Samuel Pepys and Virginia Woolf. As they recorded their own daily lives, they spoke of history too, and society. Both diaries yield insights into what it might be like, not only to live, like they did, but to live at a particular time, in a particular culture and society. Their lived experience is not only of themselves, but of themselves in a social setting, faced with particular social problems and privileges as well as personal troubles.

This tension between the private and the political is terrifyingly apparent in Anne Frank's *Diary of a Young Girl*. She begins her diary on 12 June 1942, hoping that 'I will be able to confide everything to you, as I have never been able to confide in anyone', that 'you will be great a source of comfort and support' (1997, p. 1). It is impossible not to be struck by the maturity, humour and elegance with which she shares her fears and feelings while living under prison-like conditions in a secret annex to hide from the Nazi occupiers of Amsterdam during the Second World War.

SUMMARY: FROM THE VISCERAL TO THE VISIBLE

Having discussed the various ways in which visual and biographical methods can be utilized to keep records of the embodied lives and activities that we study, it would be absurd to maintain a dualism between the kind of material which can be recorded through biographical and visual methods. Few diaries and memoirs are completely turned inwards, to only record visceral and emotional experiences we have through our bodies. Diaries may even speak about the outside world which triggered these feelings and experiences in the first place.[7]

Conversely, video and photography do not simply enable us to document and share images of what bodies look like, their movement and behaviours, gestures and body language. Visual records may even trigger visceral and emotional experiences, and they may be used to illustrate how people experience a particular issue or problem, a social situation or a working condition. The key is that the verbal and visual records that are generated with the help of biographical and visual methods offer a rich view into the empirical vicissitudes of embodied life. Once we have generated such records, we can begin to analyse them, and see how they speak to the questions and concerns that led us to generate these records in the first place. This is the main concern in our next chapter.

Questions and exercises

1. Take a photograph of something which illustrates an issue or a problem that you are currently researching. Look at the photograph and do the exercise Howard Becker borrowed from Philip Perkis. That is, take a photo, and look at it actively for two minutes (don't just stare); name everything you see, and continue to do so for five minutes; then tell yourself a story about the people and things you see in the picture; now ask yourself what emotion and mood the picture has evoked.
2. Get someone to take a photograph of you, when you are doing something which you do on a regular basis – e.g. eating, reading or writing. Again, look at the photograph and repeat Philip Perkis's exercise.
3. Have you learned something from the photographs which you would not have learned without having a photograph at hand? What have you learned? In what way do these photographs constitute what Becker calls 'specified generalizations'?
4. Keep a structured diary for one week. Make sure to record some of the bodily activities that you do (e.g. eating, writing, walking to work) and record the visceral feelings and sensations that you experience while you are engaged in this activity (you may need to write down your observations after you have completed the activity).
5. Keep an unstructured diary for a week to see how you may capture your life as you live it.
6. Expand your unstructured diary relentlessly adopting Virginia Woolf's view that introspection should give way to observation. Do this for a day, a few days, a week, or longer. How did this affect the records of your research project?
7. Compare and contrast your two diaries. Which diary offers most detail about specific experiences and activities? Which diary speaks most strongly about the main concerns in your current research? Discuss why you think this might be.

7
EMBODIED ANALYSIS AND INTERPRETATION

DISEMBODIED DECEPTIONS IN ANALYTICAL METHOD

From everyday speech and popular discourse one might get the impression that analysis is about structure, simplicity and detachment, of understanding the simple order of things, in detail, through detached logical reasoning, and without letting one's judgement be corrupted by personal interests, private passions or bodily desires. We no doubt get a hint of this when doing a simple internet search for the term, which is what many of us would routinely do in order to acquaint ourselves with a new research topic. Google tells us that analysis is 'the detailed examination of the elements or structure of something', or 'the process of separating something into its constituent elements'. In a similar vein, the keen amateurs and scholars of the Wikipedia community seem to agree that 'Analysis is the process of breaking a complex topic or substance into smaller parts in order to gain a better understanding of it.'

Given these definitions, perhaps we should not be too surprised when analysis is seen as the work of 'analytical thinkers', what some popular psychologists call 'reserved, quiet persons' who 'like to get to the bottom of things' and 'loathe contradictions and illogicalness' (ipersonic, 2017), or when organizations set up entire 'analytics teams' dominated by experts in statistical analysis, data management and computerized 'analytics tools'.[1] According to the Myers–Briggs personality test, which despite dubious academic merit remains in widespread use by recruitment agencies and human resource professionals, 'thinking' personality types are very different from 'feeling' personality types: 'thinkers' make decisions through logical reasoning; 'feelers' do so based on personal values (see, e.g., Myers and Myers, 1995 [1980]).

Although analysis may be a matter of grasping things in their nitty-gritty detail, of understanding each detail and how they are related, it also becomes a matter of simplifying the thing, problem or phenomenon that we are faced with, of reducing its complexity by distinguishing the parts from the whole, and differentiating between the parts that make up the whole. According to the late Bob Cooper (1986, in Chia and Kallinikos, 1998), these are fundamental acts of ordering; by simplifying things and putting them in their place, we tidy up the clutter that first confronts us; containing bits and pieces within neat boundaries – what Whitehead 1967 [1925] called 'simple locations' – they are no longer dirty, filthy 'matter out of place' (Douglas, 2001 [1966], pp. 36, 41) but things put in their proper place. Perhaps there is no coincidence, then, that many of our colleagues insist that they need to 'clean their data' before they can analyse them. Our worry is that such cleaning rids the empirical material of its flesh, life and complexity.

Like most inventions of western intellectual thought, analysis has its roots in Ancient Greek philosophy. After Plato in his early dialogues had set out a method of 'elenctic' analysis where the essence of a thing and the real truth of its definition was pursued through cross-examination,[2] his later dialogues unfold an analytical method of 'hypothesis', which more clearly presages the notions of analysis that we have mentioned above.[3] While hypothetical analysis is conducted by confirming a true belief with a true account, it also involves collection and division, that is, collecting different kinds of things which have something in common and then dividing them by specifying how they differ. In *The Statesman*, Plato argues that the king, the household manager and the statesman all yield their power to rule others by means of their understanding rather than their bodies (259c), but that they differ in terms of what kind of knowledge they possess as the king lacks expertise in the realm of herds (287c). Since their overlapping expertise means that they cannot be divided through dichotomy, Plato proposes an analytical 'division by limbs': that their specific expertise can only be found by dividing them 'limb by limb, like a sacrificial animal' (287c) being slaughtered and dismembered.[4]

Aristotle, who founded the subject of formal logic and whose influence on western science for a long time exceeded Plato's, may have done more to dismember and disembody the method of analysis than his former teacher. Distanced and detached from the kind of lively arguments we find in Plato's dialogues, Aristotle's writing in the *Prior* and *Posterior Analytics* is abstract and formulaic. Critical of Plato's method of division, Aristotle returns to the teachings of Ancient Greek geometry. According to the philosopher Patrick Byrne (1997), it is via geometry that Aristotle comes to define analysis as a process of knowing 'the reasoned facts' rather than just 'mere facts' (Byrne, 1997, p. 25). And, as Aristotle (340BC) confirms in the *Nichomachean Ethics* (Book III, 3), it is geometry which inspires his syllogistic analytics, where the elements of arguments are elaborated, structured and reduced to another, and where the causes of effects are found by working backwards: from ends to means; from effect to cause; from what we seek to know to something that we already know (Beaney, 2017).

The emphasis on decomposition and regression which we find in Plato's and Aristotle's analytical methods is extended in early modern notions of analysis. In *The Great Instauration*, Francis Bacon (1858 [1620], p. 25) insisted that 'the sciences need ... a form of induction which shall analyse experience and take it to pieces'; in 'The Objections and Replies' to his *Philosophical Meditations*, Descartes (1984 [1642], p. 110, §4.2) argued that 'Analysis shows the true way by ... attend[ing] even to the smallest point'; and in the *Opticks*, Newton (1730, p. 405) stressed that 'the Method of Analysis ... proceed[s] from Compounds to Ingredients, ... from Effects to their Causes'. At around the same time as Newton, Leibniz had pioneered a method of analysis whereby proofs could be made objectively, without relying on intuition, by reducing a proposition 'into simpler ideas and simpler truths' (1989, p. 217; see also Look, 2013).

Much later, towards the mid-20th century, when the social sciences were being established as distinct academic disciplines, a decompositional form of analysis was picked up perhaps most famously by Talcott Parsons (1937). In *The Structure of Social Action*, he appeals to what he calls 'analytical realism', which he distinguishes from 'empirical realism'. According to Parsons, it is by cultivating an analytical realism that he is able to develop concepts, which 'correspond, not to concrete phenomena, but to *elements* in them which are *analytically separable* from other elements' (p. 730, in Hedström, 2005, p. 3; our emphasis). In contemporary sociology, Parsons' legacy has been continued by researchers working under the heading of 'analytical sociology'. According to one of its champions, Peter Hedström (2005), the aim of analytical sociology is to identify explanatory mechanisms by 'dissecting ... social phenomena' and 'decompos[ing] a complex reality into its constituent elements' (p. 2). In our own field of organization studies, an equally anatomical logic is apparent in the tendency to analytically dissect and cut out organizations from their outside environment, and in the widespread use of body part metaphors such as 'head office', 'foot soldiers' and 'the cultural heart of the organization' (Dale, 2001).[5]

THE CORPOREAL SCARS AND PORES OF ANALYTICAL WORK

That analytical sociology views itself as a practice of dissection is not entirely accidental. During the French Renaissance the Latinized term 'analysis' did not merely become a matter of 'unravelling' and 'investigating', but synonymous with 'dissection' and 'cutting up'. Before Descartes had established his rationalist form of decompositional analysis and pioneered the shift from medieval and Renaissance scholasticism to early modern philosophy, he had developed an interest in anatomical dissection. After he had left Paris for Amsterdam in 1628, he lived for a while on Kalverstraat in Amsterdam's slaughterhouse quarter in order to secure 'a fresh supply of organs to dissect from the butchers' (Rodis-Lewis, 1992, p. 53), optimistic that the anatomical structure and mechanistic physiology of animals would parallel that of human beings.[6] Dissection, then, was not merely a metaphor of analysis, but a concrete work process that made it possible to identify and locate the different organs by which the organism was constituted.[7]

However, when tracing the etymology of analysis back to its Ancient Greek roots we are reminded that it cannot be limited to the practice of finding and fixing a thing's or a body's mechanistic structure. For the Ancient Greeks, *analuein* was literally a question of cutting (*luein*) up (*ana*), of solving and dissolving the whole and cutting its members loose rather than dissecting, dismembering and decomposing the whole into simple components with fixed locations (see Byrne, 1997). Solving problems through analysis did not simply rest on the mind's difference and division from the body, as Descartes had come to assume nearly two millennia later. As we saw in Chapter 1, Aristotle diverted from the mind/body dualism which Plato had insisted on at the time and acknowledged that the development of scientific knowledge depended on the body's capacity for sensory perception.

According to the neuroscientist Antonio Damasio (1994), Descartes' error consisted in dividing the mind from the body, thus ignoring how much our capacity for rational and analytical thinking, decision-making and problem-solving depends on our embodied feelings and emotions. Through a number of experimental studies, Damasio has shown that neurological patients who lack the capacity to feel pleasure and pain also have an impaired capacity for rational, let alone ethical, decision-making because their loss of feeling makes them stop caring about the consequences of their decisions.[8] Over the past couple of decades Damasio's pioneering work has coincided with several contributions in neuroscience and psychology which cut loose the practice of analysis and analytical thinking from the mind/body dualism and acknowledge how closely tied up it is with our feelings and intuitions (see, e.g., Pennycook et al., 2015).[9]

These insights make us realize that the problems, practices and possibilities of analysis are wide open, and that they always have been. In the *Mathematical Collection*, the Ancient mathematician Pappus of Alexandria distinguished problematical analysis from theoretical analysis: while theoretical analysis demonstrates through theorems that something does or does not exist, problematical analysis 'serves to carry out what [is] desired to do' and create something which does not yet exist (in Hintikka and Remes, 1974, p. 9).

In the second half of the 20th century, the French philosopher Gilles Deleuze cultivated such a desire through a series of unconventional readings of key figures in the history of philosophy, such as Hume, Kant and Bergson (see Deleuze, 1983 [1963], 1988 [1968], 1991 [1953]). In Deleuze's own words,

> I coped with ... the history of philosophy as a kind of buggery or ... immaculate conception. I saw myself as taking an author from behind and giving him a child that would be his own offspring, yet monstrous. It was really important for it to be his own child,

> because the author had to actually say all I had him saying. But the child was bound to be monstrous too, because it resulted from all sorts of shifting, slipping, dislocations, and hidden emissions that I really enjoyed. (Deleuze, 1995 [1990], p. 6)

Although the concerns of Pappus, Deleuze and the history of philosophy may seem far removed from the immediate concerns of most social scientists, they enable us to rethink what analysis is and what it can be. Approaching analysis as an act of buggery, we recognize that the history and future of analysis is one of pores and openings as well as wounds and scars.

ANALYTICAL BUGGERY

By 'buggering' analysis, we are able to cut it loose from the distanced, detached and desensitized manner in which disembodied researchers tend to dissect and stitch up social life as if it were a lifeless corpse. Far from being an invitation to 'make things up' or abstract things beyond recognition, it enables us to *embody* the work of analysis; to express the vibrant, untidy and fleshed relations through which we live in the social world whilst searching for the prevailing patterns and aberrant practices which shape our lives.

The practice of analytical buggery is both a theoretical project and a problematical one, which defies deductive and inductive forms of logic. Analytical buggery neither applies existing theory deductively as a grid of concepts to be confirmed by empirical data, nor does it create new theoretical constructs inductively by solely looking to what is in 'the data'. No monstrous offspring can be conceived, and no analytical buggery can be committed if one's primary concern is to see how well 'the data' fit existing theory, or if one claims to be unaffected by existing concepts and theories.

Furthermore, analytical buggery transgresses the abductive and iterative kinds of reasoning that have grown popular amongst social scientists in recent decades. Like many of our colleagues we have sought to analyse and theorize our empirical material 'abductively' and 'iteratively', by working back and forth between the material and our research problems, to see how the material answers our questions, and how far it agrees or disagrees with existing concepts and theory. In hindsight, we may argue that this is how Torkild and his co-author Louise Wallenberg came up with the notion that male-to-female transvestites express transgender by *underdoing* gender: that transvestites express transgender not by copying feminine stereotypes to pass as women (as West and Zimmerman's [1987] concept of 'doing gender' might suggest) or by exaggerating such stereotype through drag (as Butler's [1990, 1993] theory of 'gender performativity' implies), but by combining a variety of 'feminine, masculine and ungendered practices and attributes' (Thanem and Wallenberg, 2016, p. 266).[10]

While the concept of 'underdoing gender' could be seen as a positively monstrous creation which both is and is not an offspring of Judith Butler, we should stress that there is more to analytical buggery than the iterative and abductive movement between theory and empirics. If analytical buggery is to play a role in advancing the practice of embodied research methods, we even need to bugger the impersonal and disembodied approach with which Deleuze encounters it. In addition to enjoy the kind of slips and dislocations that Deleuze made whilst buggering key figures in the history of philosophy, we need to acknowledge how our own embodied experiences seep into our analytical processes.

This has always been an important undercurrent in feminist and postcolonial scholarship, though it has not always been expressed explicitly and straightforwardly. As we revisited Frantz Fanon's (2008 [1952]) anti-racist manifesto *Black Skin, White Masks*, we were reminded that his analysis of race relations and his argument for black liberation must be read bearing in mind that he 'was born in the Antilles' and not in Africa (p. 14), but also in light of the personal experiences that he shares of everyday racism in postwar France. In a more explicit way, Adrienne Rich (1976) tells us that her feminist analysis of motherhood 'is rooted in my own past' and 'that only the willingness to share private and sometimes painful experience can enable women to create a collective description of the world which will be truly ours' (pp. 15–16).[11] Interestingly, that same year Simone de Beauvoir (1995 [1976]) admitted in an interview (25 years after *The Second Sex* was first published in France) that her analysis of woman's emancipation, liberation and transcendence, and 'the belief that man and woman could be equal', was 'reinforced' by her own success as an intellectual.[12]

Though opening ourselves to accusations of post-hoc rationalization, we would be lying if we did not recognize how our own work keeps getting buggered by our own lived experiences. In Torkild's case, this is apparent in his research on transgender:

> I'm not sure what Louise and my analysis of how transvestites underdo gender would have looked like without my own experience of living as a part-time transvestite. In some ways, I was living that underdoing too, you know, using too much make-up but not giving up my masculine voice. So even if I didn't recognize myself in everything that our participants told us – for one thing my experience has been rather more joyful than much of what most of them shared with us – I recognized myself in much of the underdoing in the combination of masculine, feminine and ungendered practices.

For David, his lived experiences are particularly profound in his research on masculinity:

> I am unable to separate out my continued interest in how masculinities and gender are done from my own experience of having been infertile within a conventional heterosexual relationship, and underplaying, dissembling or even denying the existential feelings of inadequacy that were always present and liable to explode. This event had a very

> significant and somewhat explosive 'out of body' experience when I had what is known technically as depersonalization where the mind leaves the body and leads to or reflects an intense anxiety and fear that there will be no return. Perhaps this is the ultimate form of disembodiment though in feeling like a devil high above in the skies towering over my body yet peering into my very soul and providing no prospect of ever returning to the body that I had previously inhabited. However, because of being absent from my body it was difficult to feel anything other than fear. (Strangely I had a not dissimilar floating away from my body event through meditation some years earlier but this was wholly serene leaving me completely calm yet elated and alive.)
>
> Some few days after the negative experience of leaving my body, I had a nervous breakdown which was controlled with drugs that I took for seven weeks before going 'cold turkey', which was even worse than the depersonalization experience. This involved two days in which I had the sensation of my body being dragged into the depths of the earth and taking my mind with it – ironically the exact inverse of my previous 'out of body' experiences. However, this desperate and most horrifying of acts of an immediate and total withdrawal from the drugs took me out of my breakdown to which fortunately I have not returned during the rest of my life. Nonetheless, it took some time for me to realize that one of the strongest triggers for this negative episode in my life was having intellectually but not emotionally and bodily confronted and dealt with my infertility. Because I had written on masculinity, I was sure I had come to terms with it for myself but clearly not and eventually my world came crashing down on me to push my body and mind to a very 'dark' place.

Deleuze (1995) speaks about such an experience in his reflections on buggering earlier philosophers but more positively in relation to Nietzsche whose work can lead individuals 'through the harshest exercise in depersonalization, by opening themselves up to the multiplicities everywhere within them, to the intensities running through them' (p. 6). This is not to say that analytical buggery is a purely subjective exercise where our analysis is wholly determined and legitimized by our own personal experience, or that it enables us to fabricate fake facts and untruths and pass them off as scholarship. What we are arguing is that there is a personal and embodied dimension to any analysis, including our own attempts at analytical buggery.

Thus, there is an interpretive dimension to analytical buggery, which is not purely textual. Neither analysis nor interpretation can be conducted in isolation, as if delimited by the boundaries of our empirical material, our initial research problem and our literature. As the etymology suggests, interpretation must happen between things – in our view, between 'these things' and 'our embodied experiences'. The themes we identify and our interpretation of them are infused by our own lives; by our personal thrills and troubles as well as by our research problems and the research literature. Like any text, the transcripts of our empirical material make up a *corpus*, a body brought into this world by other bodies.[13]

HANDS-ON RATHER THAN ON-SCREEN

To bugger an analysis it may be good to first get a physical grasp of your empirical *corpus*. This may sound like an analogy, but we mean it literally. Forget about all the software packages that are advertised on the premise that they would do the analysis for you. Instead, rediscover some of the old unspoken tricks that used to be common practice among qualitative scholars. Get your field notes out. Type up your interview recordings and print them out along with any photographs that you have taken and any documents that you have collected. Read through your hard copies, take notes in the margins, and highlight key words and expressive quotes. Use different colours if you like. Mark important pages with stickies and write key words on catalogue cards. Type up or photocopy notes pages containing key quotes and observations. Take a pair of scissors and cut out separate quotes and observations. Lay it all out on the floor. Add artefacts that you have collected during the research process. Step into your empirical material, get on all fours and work with it, play with it. Arrange it according to themes and sub-themes without getting stuck in one structure. Draw maps and diagrams, and take photographs so that you can arrange and rearrange your material whilst keeping track of the different linear and non-linear connections that start emerging.

This may seem terribly old-fashioned at a time when the digitalized economy is so hyped up that we are constantly encouraged to read everything on-screen rather than as hard copy (for alleged environmental reasons or because of the hassle of carrying books on planes and commuter trains or the difficulty storing print-outs in hot-desked office spaces). However, recent research in pedagogy suggests that people understand and comprehend text better when reading it on paper than on-screen; because the radiation from screens undermines our ability to focus our eyes, and because it is easier to work actively with printed text (Mangen et al., 2013). As you move and shift, switch and dislocate the different bits and pieces on the floor, you may start to notice expressions and emissions in your material that were initially hidden; what you first expected to be a marginal issue may turn into a central theme, and an extraordinary event may become more significant than a common everyday practice.

Without getting into too much detail, let us offer a couple of brief illustrations from our own experiences of working with this hands-on process. From Torkild first:

> In my work with Louise on transgender, it was only after having spread out our highlighted transcripts, post-it notes and clippings on the floor that we started to realize what 'underdoing gender' might involve; it was only then that we recognized that our participants combined feminine dress and masculine voice with a kind of self-esteem, confidence and assertiveness; it was only then that we came to appreciate that they did so through a variety of bodily postures, facial expressions and pedestrian styles, which cut across gender stereotypes and binaries, and which required a kind of energy and determination that has nothing to do with gender. While this struck a chord with my own experiences of crossdressing in public, it all somehow came together when we mapped it out on the floor.

David's experience is more 'on-screen' yet 'hands-on' nonetheless:

> Having done far more research and writing with others than on my own I can honestly say that analysing research material and composing the words and sentences for a book or paper has worked more effectively when I've sat down with my co-author in the same room than when we've emailed text back and forth. In some of my very early work on masculinity with Deborah Kerfoot we even shared the same keyboard. Our conceptualizations of masculinity in management taking the form of conquest, competition and control were generated out of this interactive process (Kerfoot and Knights, 1994). I've kept up this habit in my work with Caroline Clarke where we have developed ideas about the troublesome concept of 'perfection' that sometimes renders academics and veterinary surgeons debilitated by self-blame when their writing/teaching or animal cures fail to meet an ideal standard. This, we argue, is linked largely to their training and the requirement to secure high-level qualifications before practising their profession (Clarke and Knights, 2018; Knights and Clarke, 2018), such that success rather than failure is the norm. Partly because of living in different countries and partly because of my heart condition occurring in the middle of this book project, this has been more difficult to do with Torkild.

SUMMARY: FROM DISSECTION TO BUGGERY

Closing this chapter, let us invoke a note that Henry Thoreau made in his journal on 7 March 1838: that 'We should not endavor [sic] coolly to analyse our thoughts, but, keeping the pen even and parallel with the current, make an accurate transcript of them. Impulse is, after all, the best linguist' (Cramer, 2007, p. 4). Although we do not endorse everything in Thoreau's claim, in this chapter we have sought to articulate a kind of analytical buggery which offers an alternative to the disembodied, detached and decompositional notions of analysis prevailing in scholarly and lay discourse.

Bearing in mind the scars and openings that are embodied in historical and present writings on analytical method, analytical buggery makes it possible to untangle our research problems by infusing new sensations, extracting hidden emissions and creating unconventional insights from the empirical material at hand. By confronting our empirical material with our own embodied experiences as well as with existing theoretical concepts, analytical buggery makes it slip out of the established frames of reference that otherwise tend to determine our thinking. Cutting our empirical material loose from established concepts, analytical buggery may create immaculately lucid concepts. Though such concepts are unable to represent the empirical reality we enquire into, they may express distinct features in the lives of our research participants without neglecting their contradictory experiences and indeterminate vicissitudes. Unlike the dead corpse facing the dissector, analytical buggery recognizes that empirical matter is living, breathing and thinking flesh.

Questions and exercises

1. The aim of this exercise is to compare doing a conventional analysis with analytical buggery.

    a. Take an interview transcript, an extract from your field notes or another empirical text and subject it to a conventional, decompositional analysis. Your aim is to reduce the transcript to its most simple elements without letting the exercise be affected by bodily impulse or personal experience. List the elements resulting from your analysis and see how they might connect by ordering them hierarchically according to themes and sub-themes. Feel free to iterate between the transcript, your research problem and extant concepts in relevant literature.

    b. Now, take the same transcript and subject it to analytical buggery. What themes and connections emerge? In what way are they similar or different from what you produced through your conventional analysis? If the result of your buggered analysis is little different from your first analysis, go back and bugger it more. How much embodied impulse or personal experience is expressed in your buggered analysis? How, and how much, does it change the themes and components in your analysis?

8 EMBODIED WRITING AND DISSEMINATION

WRITING WITH OUR BODIES

In *The Sociological Imagination*, the *enfant terrible* of 1950s sociology C. Wright Mills (1959, pp. 25–32) has a go at translating Talcott Parsons' (1951) book *The Social System* into plain English. With poorly hidden sarcasm, Mills argues that Parsons' book of 555 pages can be condensed into little more than 150 pages, without losing any of its substance yet gaining in readability. Parsons was widely regarded as the most influential sociologist at the time but, according to Mills, his 'grand theory' simply covers up old problems in abstract concepts; Parsons' many long and difficult sentences do nothing to improve our understanding of empirical phenomena in the social world.

As Parsons' status declined, Mills' influence increased. In the early pages of his book *Writing for Social Scientists*, Howard Becker (2007 [1986]) tells us how he had Mills' exercise 'in mind' when teaching writing to a class of sociology students at Northwestern University. Working on texts that his students had brought to class, they 'changed passive to active sentences' (p. 6), removed unnecessary words and phrases, replaced jargon with ordinary words, broke up and shortened long sentences: 'At the end of three hours, we had reduced four pages to three-quarters of a page without losing any nuance or essential detail' (p. 6).

Neither Mills nor Becker have been explicitly concerned with the problems of embodied research, but they have likely influenced what social scientists judge to be good academic writing.[1] Fewer of us seem to take pride in writing long passive sentences, filled to the brim with technical jargon and muddled abstractions. Fewer of us seem to hide behind a 'façade of impersonality' (Becker, 2007, p. 15) where we pretend that we are not part of the texts we write, or fantasize that we can present our findings and arguments objectively and neutrally as long as our prose remains passive and detached. Much academic writing has become much less detached and less disembodied, more active and direct, more engaged, more engaging and more readable.

At the risk of being hoisted by our own petard, we will in this chapter try and build on these advancements to flesh out our appeal to embodied ways of writing. While no literature captures life, some styles of writing can more vividly express the lives we live. This is not just a matter of writing about our own bodies, or viewing writing as an instrument, however vital, of embodied research; one of the key points that we have pursued in this book is that embodied research connects our lives and troubles to those of other people. Since writing shapes how we think about our research and how we communicate it, our bodily connections need to be reflected in our writing too, even in the texts we publish. Writing can never be an isolated stage of 'writing up'.

In the first part of this chapter, we shall discuss some previous efforts to craft more embodied ways of writing, especially the 'feminine writing' which springs out of French feminist philosophy and literature. Later, we shall address some of the physical aspects of writing, including some of the anxieties and pains that are related to writing and writer's block as well as some of the routines that writers use to handle these problems. Finally, we shall discuss some of the challenges of disseminating, publishing and presenting embodied research, recognizing that the term 'dissemination' itself is laden with malestream connotations.

FEMININE WRITING AND PHALLOGOCENTRIC WRITING

In previous chapters we have strived to flesh out various aspects of the embodied research process through examples that describe social life with vivacity, rich nuance and intensity. An important tradition in this respect is the *écriture féminine*, or 'feminine writing', which was pioneered by the French feminist philosophers Hélène Cixous (who first used the term) and Luce Irigaray. In her breakthrough text 'The Laugh of Medusa', Cixous (1976 [1975]) argues that 'Woman must write her self: must write about women and bring women to writing, from which they have been driven away as violently as from their bodies' (p. 875).

Though we sympathize with Cixous' appeal, it is not readily apparent what it might mean for male academics, or how men in the social sciences may take up her challenge. Whereas David has drawn on Cixous and Irigaray in his joint work on masculinity with Deborah Kerfoot (see Kerfoot and Knights, 1994), Torkild has discussed Irigaray's (1985 [1977]) critique of phallogocentrism in a piece on Swedish sex education (Thanem, 2010). But when drafting this chapter in January 2018, Torkild in particular found himself frustrated by feminine writing, worrying that he is incapable of engaging with it in a genuine manner:

At this stage, I'm annoyed and exhausted; by this book and this chapter, by feminine writing and by my own writing. My head aches, and my shoulders ache. I have a sore throat. In our proposal to the publisher, which we sent off in February 2014, we promised that we would have the book done by June 2015. We were optimistic, stupidly optimistic, and gravely underestimated how much work would be required, at least if we were to do this properly. It's now January 2018, and I keep telling myself that there are only two more chapters left to write, including this one, but not counting the conclusion, which I convince myself will be short.

I find it difficult to write about feminine writing; to write about it in a way which doesn't do too much violence to it, which doesn't co-opt it for my own selfish purposes by squeezing it into the straightjacket of the phallogocentric tradition which it writes against. While David and I have long regarded ourselves as supportive of feminine writing and its politics, there's no denying that we are biological men, with quite different desires experiences and privileges.

I mention my qualms and frustrations to Louise as we take the dog for a walk. The flowery prose, the non-linear and slippery reasoning, the vague and un-pointed arguments, the waves, floods and outbursts that Cixous speaks of.

'This whole thing bothers me, Louise, not least because it's got so much in common with what Gilles Deleuze was doing when he buggered the history of philosophy!'

'So why don't you write about it?' she tells me with a smirk.

'I guess I should', relieved that it might help me write the damn chapter yet anxious that I may still corrupt feminine writing into a malestream farce.

'Why do you think you're so frustrated by feminine writing? What do you think *Speculum of the Other Woman* means, by the way?'

I stare back at her, dumbfounded.

'You know, the title of Irigaray's book.'

'Well ...'

'A speculum is a medical instrument, which male gynaecologists would need in order to look into the woman's vagina. And "the other" woman is not simply "another" woman.'

'No, I do realize that', I mumble.

'When Irigaray speaks of the "other" woman she suggests that woman is an other who is capable of looking into the woman body without the assistance of gynaecological instruments, understanding and writing about women's bodies without being corrupted by phallogocentric perspective.'

As I'm instrumentally writing this up, I admit to myself that this is the problem; that I've been soaked in phallogocentric marinade for so long that I'm not capable of appreciating the beauty and sophistication that is embodied in feminine writing, or relate to the specific yet plentiful experiences that are expressed through it.

Irigaray and Cixous shed light on exactly these kinds of difficulties when drawing analogies between woman's sexuality and feminine language, reminding us that the etymology of literary 'genre' is gendered (see also Battersby, 1989). A few pages down in 'The Laugh of Medusa', Cixous likens woman's sexuality to 'her writing', suggesting that both are 'cosmic, not centred on one organ' (1976, p. 889). In *This Sex Which Is Not One*, Irigaray (1985) argues that 'woman has sex organs more or less everywhere' (p. 28) and that, consequently, 'her language ... sets off in all directions leaving "him" unable to discern the coherence of any meaning' (p. 29). For Cixous, experimental writing is conditioned on erotic experimentation. For Irigaray, feminine language is enabled by the essential fact that woman's sexuality and pleasure is self-contained, as the self-touching labia require no external member, instrument or body to please themselves.

I hesitate writing this down, unsure that I can write it without coming across as a peeping Tom. It feels unlikely that the multiplicity of woman's desire, sexuality and anatomy that Irigaray speaks of is embodied in my own transvestism, which depends on the kind of external artefacts and instruments that no woman needs. But as it feels safer to write as a thief than as a pervert I can readily admit that I steal shamelessly from feminine writing, especially Cixous' point that, to write, you need to write your self and your body, even though it scares and upsets men, or rather, I tell myself, because it does. Cixous refuses to accept that writing is reserved for 'Great Men' (1976, p. 876) and launches instead an 'inevitable struggle against conventional man' (p. 875). Thus, writing is not merely a tool for defining or illuminating the research problems we have investigated or showing how social problems are connected to personal troubles. Feminine writing makes trouble. It makes gender trouble, and it does so in quite a different way from Judith Butler (1990), whose argument is rigorous but abstract.

We can hardly claim to be feminine writers. As journal editors, reviewers and co-authors, we have oftentimes subjected other people's writing, some of it explicitly feminine, to our phallogocentric preference for ostensible clarity, conciseness and linearity. And in this book, many of the exemplar accounts we have drawn on from other people's work have not been selected for their poetic qualities, but because they describe feelings, emotions, gestures and practices that real people express and experience through their bodies. Indeed, they do so in more obvious ways and in more straightforward language than some of the feminine writing.

Loïc Wacquant, for one, writes beautifully, accessibly and with apparent ease, though his writing is sometimes unreflectively masculine. Reading his 'sociological novel' *The Body and Soul* we are struck by the vibrant intensity with which he describes his own visceral feelings and experiences whilst investigating one particular corner of the male-dominated world of boxing. We find similar qualities in C. Wright Mills' writing, in spite of the macho style, which was not unparalleled by a life of excessive smoking, heavy drinking and fast motorcycles (see, e.g., Horowitz, 1983; Wakefield, 2009); when Mills 'edited' Parsons' 'grand theory', he pulled down Parsons' trousers, thus revealing an emperor without clothes.

So, where does this leave us? Being both troubled and seduced by feminine writing and by its macho adversary, perhaps our own writing is little but a buggered assemblage of both.

WRITING IS A PHYSICAL ACTIVITY

A recurring theme in this book concerns how research and writing is connected to the researcher's own body, embodied identity and bodily experiences. However, writing is also a physical activity in itself, in a more concrete and mundane sense than we have acknowledged thus far. Like Becker (2007), we are particularly reminded of 'how physical writing is' (p. 152) when we succumb to writing in long 'marathon spurts', sitting for hours by our desks or in an armchair with the laptop, hardly getting up to eat, drink or go to the bathroom. We know perfectly well how this sedentary habit translates into a pain which spreads from the neck, down through our shoulders and back, and out to our arms and legs. We know equally well that our constant staring at the computer screen makes our eyes sore and red, that it strains them to the point of causing involuntary spasms and pulsations.

Much of this has to do with writer's block, and the fear of getting stuck in it. It is not easy to follow Becker's advice and just treat the first piece of writing as a rough draft. It is just as difficult to follow the advice we sometimes give our own students: to get it written before you worry about getting it right. Torkild can relete to this:

> I don't know how many times I've pulled my hair, and kept fooling myself that I'll get up and take the dog for a walk, put the kettle on, make lunch or cook dinner, just as soon as I've gotten that sentence right, finished that bloody paragraph, or integrated all my 'darling' ideas into the text, panicking, that if I don't, I won't have enough to fill the page.

On top of this, there are all sorts of distractions, particularly, as David admits, when working from home, as there are always jobs to do around the house or in the garden in addition to constantly consulting the emails and the internet. It is much easier to check the email or read the news online or do whatever, instead of getting the writing done. As psychologists are currently exploring the neurological causes and reward mechanisms of excessive emailing, instant messaging and internet use (e.g. Montag et al., 2015, 2016), we are painfully aware that the rewards one might get from writing long pieces of research are anything but immediate. While working to deadlines may not be wholly devastating but push us to write, there is more help available for those who want it, and it does not need to come in the shape of a pill filled with amphetamines. Academics and other writers have developed a number of more or less conscious strategies to get their writing done (or to avoid writing).

As Becker (2007) has mentioned, some people have a 'physical addiction' to particular pens, notepads and other writing tools. They cannot write without them. Others write best in a particular physical environment. Most people need peace and quiet to be able to write, not to be disturbed by chatter, telephones and music. While your office or study will hopefully provide such an environment, some people prefer to write in other spaces, at least some of the time. Many people find it hard to get any serious writing done in their office because of interruptions from colleagues and students. Others may find it difficult to resist the many distractions of the home environment, telling themselves that they will start writing just as soon as they have done the dishes and the laundry, vacuumed the floors, alphabetized their books, organized their sock drawer, done the recycling, watered the plants and mowed the lawn. Yet others go on writing retreats to cabins and holiday resorts in remote locations, only to find themselves consumed by the beautiful scenery or what else the place has to offer. Sometimes it may actually be easier to break the monotony and stay focused on your writing if you sit yourself down at the library, in a bookstore or in a nearby cafe. To get any writing done on the weekend, Torkild has developed a habit of writing in front of the telly, half-sitting half-laying down on the sofa. Although this routine adds a recreational dimension to work, it is not something we would recommend. Better perhaps to adopt David's routine of playing classical music from the radio, as long as there is not too much singing or presenter talk, that is.

Professional writers tend to be more disciplined. They often have strict routines for when to write and how much to write every day. Don DeLillo (1993) and Stephen King (2006) both start writing in the early morning and keep going for a good few hours without a break. Alice Munro (1994) followed a similar regimen before retiring in 2013. Ernest Hemingway (1958) would start 'writing as soon after first light as possible' (he even wrote standing up, several decades before the height-adjustable desk became a standard fixture of the modern office). Maya Angelou (1990) used to isolate herself in a stripped-down rented hotel room between 7am and 2pm (if the writing went well; until 12.30pm if it did not); hotel staff were only allowed to come in when she was out to empty the waste basket. Don DeLillo finds solitude at home. Stephen King prefers a room where he can close the door and draw the curtains. After a three-hour spurt of writing Alice Munro would go for a long walk, at least three miles. After four hours Don DeLillo would go for a run.

The key thing is the writing: that is, getting down the words and sentences on the page. Editing and rewriting comes later. Alice Munro used to have a quota of pages that she pressed herself to meet every day, finding a sense of protection in the various routines and rituals. Stephen King has said that he 'used to write two thousand words a day and sometimes even more', though it has later dropped to 'a paltry thousand words a day' (2006). Nevertheless, his writing is not detached from his reading. Like most professional writers, he constantly reads other people's work to make sure that he knows what good writing reads like (King, 2000).

This all makes perfect sense. But reading other people's advice is one thing. Putting it into practice is something entirely different, as Torkild can testify.

Adopting a new habit

It's 2.30pm on 30 January 2018. I've written just over 400 words, which is an embarrassing contrast to the writing professionals above, and to the example of Anthony Giddens. When I was doing my PhD, my supervisor told me that Giddens once declined to take a late-morning walk with his colleague Margaret Archer because he hadn't done his 3000 words yet.[2]

To improve my pace and to break the frequent onset of writer's block I've ended up trying many of the tricks that have been promoted in recent years: I've tried closing down my email programme completely, only allowing myself to check emails after I've completed a couple of hours of writing; I've tried keeping the web browser off, which is difficult because I frequently need it to confirm things that I'm writing about; I've tried Pomodoro, the smartphone application which encourages users to work in sequences of 25-minute intervals followed by five-minute breaks; I've even tried to get my energy up by doing push-ups during some of the five-minute breaks (apparently, Kurt Vonnegut did 'push-ups and sit-ups all the time' when he started teaching on the University of Iowa Writers' Workshop in 1965. At least, this is what he wrote in a letter to his wife Jane dated 28th September 1965 (see Vonnegut 2012)). These tricks work until I get too bored to stay focused or start fretting too much about being consumed by the endless white space, even if it's just half a page.

It gives some comfort to learn that Margaret Atwood (2013) relies on no specific routine, place or writing tools to get her writing done. She writes a hell of a lot because she keeps writing, even when she's not in her study, even when the routine gets broken. This lesson may indeed make it easier to break bad sedate habits, to take short breaks more frequently, stand up, move about and get the bloodstream going. If not, there's always David's trick: drink loads of coffee! That'll force me to get up. To make the coffee, and go to the loo.

SHARING AND PUBLISHING EMBODIED WRITING

Even though writing can be both difficult and painful, not writing is hardly an option for contemporary social scientists. Nonetheless, it would not hurt to reflect more about why we write and how we write, how we share our writing with others and how we try and publish it.

In December 2017 Alison Pullen visited Torkild's department to give a workshop on writing. She had emailed some instructions in advance. Before meeting, she asked participants to bring some of our own writing, and to consider the following questions: 'What do you enjoy most about writing? Does anything frustrate you about writing journal articles? What is the hardest part of writing?' (Alison had even included her own answers to these questions, admitting that she most enjoyed 'writing without constraint', that she was frustrated by 'rewriting ... [her] text to align with what reviewers want to read', and that she found it hard to keep 'to one idea in one paper'.)

We rarely ask ourselves that first question: What do we most enjoy about writing? And we rarely allow ourselves to write freely. On top of the frustrations and anxieties of writer's block and the physical pain of sitting down to write hour upon hour, we impose all sorts of constraints and boundaries on our writing. Of course, we are pressured to publish by our universities, funding agencies and colleagues, but a lot of that pressure comes from ourselves and our own obsession with writing and publishing. We often think about where to publish a text early in the writing process, sometimes even before we start writing. We had hardly started writing this book before we approached the publisher with a proposal! At the same time, countless papers are written specifically for a conference stream, a journal, a special issue or edited collection.

Thus, publishers, editors and reviewers do not only affect how we rewrite our texts. As they decide whether our manuscripts are publishable or not (or how much work is required to make them publishable), they shape our writing and thinking. Like most professional readers, they tend to read and judge our writing according to the rules and style of a particular genre, though when it comes to books the main concern seems to be the market and how much potential profit the publisher might make from the contract.

This is not only a bad thing. If comments are made in a constructive manner rather than to prove us wrong or make sure that we have paid sufficient homage to established authors in the field, they may help us clarify our argument, sharpen our contribution, reach a decent-sized audience and write better! Actually, a number of journals (the *Academy of Management Journal* to name one) have developed guidelines for reviewers to prevent the kind of destructive tendencies that we mention here (not that this has prevented them from publishing a majority of articles that are filled to the brim with technical jargon and lengthy literature reviews which undermine any attempt to say anything remotely distinct or exciting).

Nevertheless, for many of us our writing is crippled and blocked by a fear of failure: that editors will reject it and that nobody will publish it, let alone read it and certainly not cite it if it finally ends up getting published. While 'the paralyzing fear of having others read your work' (Becker, 2007, p. xi) may make some people reluctant to share works-in-progress with friends and colleagues, this is not just about avoiding the kind of harsh judgement that others have subjected us to in the past. Some of us may even have internalized their criteria so deeply that we block ourselves from writing in the first place. At least then we save ourselves from being debunked by the excessively judgemental reader-editor who has come to occupy our heads.

In Torkild's case, Alison's writing workshop did something to loosen up his blockage, if only for a short while. Furthermore, it stressed upon him that writing can be a social activity as well as an intellectual and physical activity:

Dividing us into two small groups, Alison asked everybody to first take 10 minutes on our own to answer the following question:

'Why do you write?'

I'm not sure she noticed the horror in my eyes. She certainly didn't let it prevent her from continuing as planned:

'Just put it down in writing, then read it out to the others in your group. When you've all done that, work together to combine your contributions into one piece of writing.'

The instructions made me feel uneasy. They pushed me out of my comfort zone and made me lose control. But since it would be more embarrassing not to have written anything on this allegedly straightforward question, I forced myself to write, and to write fast. Thankfully, I soon started to forget that inner reader-editor who keeps telling me what I can write, what I can't write, and that I can't write at all. Even though I wrote as much for the rest of the group as I did for myself, it became less terrifying to write. And once we'd all shared our writing, it proved less difficult than I'd anticipated to combine our different scribbles into one piece of text. As we strived to become one collective author who spoke for all of us in one singular yet plural voice, we actually developed a sense of critical friendship which nurtured our affinities as well as our discords. We praised what we liked and we questioned what we didn't understand. We tweaked some of the claims, dropped others, and added new ones.

While the pressure to publish has led more academics to focus on writing co-authored journal articles, that process is rarely as social and sociable as the writing experience Torkild had during Alison's workshop. By consolidating our writing efforts with colleagues, we may think that we will get our hands on more data, that we will write more articles within a given period of time, and that we stand a better chance at getting into more prestigious journals. But the instrumentality of this approach prevents us from asking ourselves or our co-authors why we write.

In hindsight, we realize that Alison's writing workshop (and Torkild's co-authors) helped us recognize that we write for a number of different reasons: for lofty or philistine reasons or simply to pursue our own instrumental self-interests; because we enjoy writing, because it's our job to write, because it helps us think and because, on some level, we want to effect change – to how we think, how we work, how we live and how we write. Unless we write, there is very little we can say about what it's like to be an embodied creature in the current world of work and organization, and we can do even less to change that world if we stay silent.

Furthermore, we realized how difficult and frustrating it has been for us to write together. Whereas Torkild has felt that David should have worked more on the book and less on his numerous journal articles and conference papers, David has felt squeezed out:

> Torkild: Perhaps I had too strong ideas early on about what I wanted the book to be about, but I just felt that we needed to get on with the bloody writing and deal with various qualms and questions marks later. So then I cut you off from much of the writing process because I couldn't face that you kept challenging so much of what I felt were the basic premises of the book. And then I got more frustrated and more impatient because we were so far behind schedule!
>
> David: I had frustrations from the beginning because even when we shared the same physical space on a couple of occasions, we rarely worked together on the same text as had been my experience with other co-authors. I should have raised these frustrations but instead proceeded to work in the same way as Torkild but then he was only willing to incorporate small parts of what I had written. Clearly, Torkild had a stronger idea of what the book should contain and so took control and tended to exclude me from much of the earlier writing process. This in some way was fortunate once I was diagnosed with major heart problems and then admitted to hospital for open heart surgery but in the earlier period it resulted in my devoting the freed-up time to other co-authored publications that were in process and often with deadlines. We had so breached the deadlines on the book that I had stopped taking them seriously until Torkild aroused my guilt with his productivity.

Although we may have failed miserably to experiment with collaboratively embodied forms of writing when writing this book, there are increasing publishing opportunities for those who want to experiment with style and genre; to write social science texts that speak *with* our bodies as well as *of* the body. We often hear that books are better suited for this purpose than journal articles. However, even in our own field of management and organization studies, at least two special issues have been published on the theme of feminine writing (see Höpfl et al., 2015; Brewis et al., 2017), and a number of workshops and conference streams have been organized. (Just a few days before writing these lines, the journal *Organization* published an article by Alison Pullen [2018], where she playfully challenges the still prevalent assumption that writing must be stripped of flesh and embodied experience, as if it were an act of labiaplasty.) While few men are involved in these projects, the publishing industry seems to be less afraid of feminine texts than it was when Cixous (1976) wrote 'The Laugh of Medusa' (see p. 877).

BROADER IMPACT AND EMBODIED AFFECT

What complicates this is that academics are more and more expected to have impact beyond our own sub-fields, cliques and collegial circles. Getting published is only a minimum requirement in a system where an academic's worth is increasingly assessed in terms of broader 'impact' and 'relevance'. Our research needs to get read and cited, sure, but there is also pressure on us to frame it in ways that make it appetizing to the mass media and to the general public. Many journalists seem to have a preference for research findings that can be reduced to simple soundbites and tabloid headlines, and universities make this worse by employing hordes of communication officers who think they know what the press wants from us. This was certainly the case when Torkild recently attended a half-day session in media training, where the main message was 'news value': 'Research gets picked up by the media because a journalist or editor thinks it's got news value. Consequently, 700 online clicks are better than 70 clicks.'

Additional pressure may come from research participants, particularly if they have granted generous empirical access. Research participants may want to read what we write about them in jargon-free texts that do not require specialist academic expertise. And they might enjoy the attention that is created when our research gets picked up by the media, at least if the image that comes through is a positive one.

This is not all bad. It depends on the media outlet, the journalist, and how you are able to work with them. Having become accustomed to extremely little media attention, Torkild recently had quite a positive experience working with his colleague Carl Cederström on a piece aimed at a broader audience.

'The Swedish CEO who runs his company like a CrossFit gym'

As mentioned in Chapter 3, my empirical access to do ethnographic fieldwork within a sportswear company rested in large part on a promise of media attention: at our first meeting with the CEO, Carl and I told him we would be writing a popular piece about him for the *Harvard Business Review*.

Although I felt a constant if largely unspoken pressure to deliver on this promise, it wasn't until a year and a bit later that we got around to writing the piece. We commenced the process 10 minutes after boarding a train for Malmö (we were on our way to a conference), and thanks to Carl's discipline and journalistic experience we had completed a first version and sent it off to the editor by the time we arrived back in Stockholm two days later (had it been up to me the piece would probably still remain unsubmitted).

Before and after we had sent off the first version, a variety of embodied experiences ensued. At first I felt humiliated by the fact that my lacking experience in the genre left me following Carl's orders. I then worried that the editor wouldn't like the piece. After Carl and I had drastically reworked it following her recommendations (Carl was in complete agreement with her,

(Continued)

(Continued)

I was less so but kept my mouth shut), I felt disappointed when she returned the text, having removed some of my most vivid and painstakingly crafted passages (including descriptions of the CEO's posture and bodily gestures during our lunch meeting). Later on, I rejoiced when she told us how excited she was about the piece, which now included financial figures and new field observations that I'd been able to contribute. However, this joy was quickly transformed into at least one sleep-deprived night agonizing that the CEO might feel misrepresented by the article's satirical tone and feel betrayed about some of the new field details. Indeed, he did email me back after I had shared a nearly final version with him, suggesting that the text could do with being 'more nuanced and facts-based'.

Nevertheless, the whole process turned out to be quite rewarding. After some minor tweaking, the CEO was happy with the piece, and didn't ask us to remove any of the more critical points that we were making. He even got two staff members to dig out data that I'd asked for since the beginning of my study but which they had previously been reluctant to provide. Furthermore, working directly with this editor was a far better experience than I had ever had working with journal editors and reviewers: she was supportive throughout the process; all her feedback and comments were constructive, detailed and immediate; and, by acting as a co-author as well as an editor, she made the text better not worse! Best of all, the whole process ended up only taking three to four months! On 12 March 2018, it was online with the following headline: 'The Swedish CEO who runs his company like a CrossFit gym' (Cederström and Thanem, 2018). A couple of days later it had even been mentioned by the *Wall Street Journal*! Well, on one of their blogs at least.

Of course, writing short popular pieces like this has its limitations, but so do all genres. In particular, it may be difficult to express the embodied vicissitudes of social life with much richness in these kinds of texts, but it may be possible to sneak in small details of it. Moreover, texts in popular outlets may draw readers' attention to pieces of work where you are able to develop your embodied writing at more length, with more opulence, intensity and nuance.

The mass media publish a variety of writing, including columns, book reviews and reportage. Dumbed-down press releases are not a golden rule. We acknowledge that it may be difficult for academics to get the attention of newspapers and other media outlets. However, we are not certain it is more difficult if you are doing embodied research. Embodied research may be less likely to get picked up in conventional news reporting (little of it can be boiled down to 'hard facts'), but it may get into the arts and culture pages and other specialist sections.

There are already quite a few examples of this. Loïc Wacquant's (2004) 'sociological novel' on boxing was a media success which vividly expressed his own embodied immersion in the pugilistic art. Sara Ahmed's work has generated newspaper headlines for a number of years, most recently with her book *Living a Feminist Life*, which, builds on personal experience and biographical detail to articulate the challenges of being a feminist in everyday life. And as we were writing this, our colleagues Carl Cederström and Andre Spicer (2017) received a lot of good press for a book which, in diary form, traces their year-long attempt to optimize their own minds, selves and bodies through the use of popular self-help techniques.

We are not suggesting it is easy to reach out to broader audiences with embodied research. It requires relentless work, on top of the academic writing process. It may even require abiding by the established rules of disembodied popular writing before breaking them. To get journalists' attention the research may need to be turned into a spectacular story, and findings may need to be twisted into loud arguments. Being a name helps, but even big names started out small. Just as academic journals in management and organization studies can be pushed to publish feminine writing, we are optimistic that the meaning of 'news value' can be challenged or renegotiated.

Another complicating factor is that media exposure may involve radio and television interviews rather than written texts for print newspapers and magazines, a format that many academics are not used to present in. While such face-to-face encounters may be more obviously embodied than the actual process of writing, Torkild has experienced how hard it can be to answer a journalist's questions related to his research in ways which retain at least some complexity to the topic without becoming too self-conscious about what he is saying and how he is presenting himself. He has also experienced how nerve-wracking it can be to expose himself and his research participants to a potentially large audience which he lacks experience engaging with.

Whereas Cixous (1976) framed 'the torment of getting up to speak' as a primarily female experience, many of us often feel our hearts racing when we open our mouths in public, sometimes so much so that we get 'entirely lost for words' and find ourselves not so much speaking as throwing our trembling bodies forward (pp. 880–1). Although malestream norms and macho tendencies in academia and journalism mean that this may be worse for women than for men, much of it is about identity and its relationship to an unrealizable notion of perfection (Bell and King, 2010; Clarke and Knights, 2018).

Cixous' solution is to literally flesh out the embodied nature of knowledge instead of denying it. Challenging our obsession with identity and perfection, she encourages us to let our bodies vitally support the logic of our speech by letting go of ourselves and letting all of us pass into our voice. It is only when we lay ourselves bare in this manner that our 'flesh speaks true' and we physically materialize what we are thinking (p. 881). Like feminine writing, feminine presenting is with the body and of the body, as our colleague Katie Beavan (2017) has powerfully demonstrated.

By endorsing embodied forms of writing and presenting we relieve ourselves from the strive to create objectively truthful representations that correspond to reality. Life exceeds literature, and our bodies exceed the categories that we seek to understand them through. Nonetheless, we may be able to write engaged and lively accounts of the worlds and cultures we embody, narratives which express the nuances and intensities of social life, vibrate the social problems and practices of our research participants against the depths and surfaces of our own bodies and experiences, and, who knows, affect readers beyond our own academic sub-fields.

SUMMARY: TOWARDS AN EMBODIED SCRIPTOLOGY

Social scientists and literary professionals are not that different. We both write for a living. The difference is that social scientists largely take the method of writing for granted. Most journal articles contain methods sections, but, as our colleague Carl Rhodes (2018) has eloquently noted, they almost never contain discussions of the writing style that an author adopts. There is methodology, Rhodes says, but not 'scriptology'. Inspired by Rhodes' appeal for a scriptology of social science research, what we have attempted in this chapter is to sketch out the main ingredients of an *embodied* scriptology.

To do so, we have addressed the bodily pains and angsts of the writing process, and discussed the increasing opportunities that now exist to write and publish embodied research. We trust we have not squandered these opportunities in this book. Inspired and encouraged by various kinds of embodied writing (feminine and not), we have tried to deviate from the rules and practices that dominate the genre of social science methods books. With some risk of imposing our own set of rules, we have tried to share our troubles and frustrations rather than offer neat lists of recommendations and toolbox solutions, fleshing out rather than silencing how our own bodies are part of the research process. As Cixous (1976, p. 880) said: 'Censor the body and you censor breath and speech at the same time.'

Questions and exercises

To help you get the most out of this chapter, we would like you to try out two rather different writing exercises. The first exercise is meant to help you write freely. The second exercise is aimed to help you stay disciplined in your writing.

1. Inspired by our colleague Alison Pullen, we would like you to pose yourself the following questions:

   a. What do you enjoy most about writing?
   b. What frustrates you most about writing?

   c. Why do you think that is?
   d. Why do you write? Write freely about this for 10 minutes. What did this feel like?
   e. Repeat the exercise with a friend or a group of friends, and share your short pieces on writing. Then take 20–30 minutes to try and assemble your individual pieces of writing into one piece of text. What did that feel like? What did you most enjoy about this exercise? What did you least like about it? Why?

2. Take a piece of text that you are struggling to complete and set aside three hours when you can work on it undisturbed. Try and have a clear idea of what you want to achieve with this text. Make sure to switch off your mobile phone and email, and only allow yourself to use the internet to search for things that you immediately need to complete a particular sentence or paragraph. Now, instead of sitting down non-stop, try if you can work by the rules of 'workflow', using an alarm clock or an application on your smartphone:

   a. Sitting down, ready to write, start the clock and write for 25 minutes.
   b. After 25 minutes the alarm will beep. Take a five-minute break. Stand up, look out the window and do not think about your writing. Do not use the telephone, email or the internet.
   c. After five minutes the alarm will beep again. Reset the alarm for 25 minutes, sit down and resume writing.
   d. Make sure to stick to the times (write when you are supposed to write and take breaks when you are supposed to), and try to keep going like this for three hours.
   e. How did you find this exercise? How much writing did you produce?

9

TRUTH AND POLITICS IN EMBODIED RESEARCH

The truths and politics of embodied research seem more contested now than they have been for a long time. Over the past few years, the body has widely come to be appreciated as a source of knowledge and as a subject of political engagement and economic activity. At the same time, the rise of right-wing populism and religious extremism is expressing the ugly face of an embodied politics, which violates human rights and robs people of their livelihood, destroys the natural environment yet denies the human causes of climate change, and commits horrific war crimes but lies blatantly even when faced with obvious facts. Without question, this is a politics which thrives on sad affects and passions; on hatred and violence, greed and arrogance, an immunity to facts, and a short-circuiting of reason and rationality.

To combat these attacks on truth and justice, several academic and public commentators seem to imply that we need less embodiment, not more; that liberal democracy must reinvent the universal rights and duties of citizenship (Lilla, 2017); that a collective class politics requires us to rid ourselves of our specific identities and 'become universal in ourselves' (Zizek, 2017, p. 181); and that bodily concerns put us on a slippery slope towards an identity politics which is too personal and which divides people according to gender, ethnicity, sexuality and other bodily features (e.g. Michaels, 2006; Smith, 2008; Žižek, 2017; Fukuyama, 2018).

We agree that a reasonable politics can hardly be based on the self-centred experiences and egoistic interests of individuals and identity groups. But is the only viable alternative one of disembodiment? Do we need to disembody the scholarly pursuit of truth and justice in order to fight the reactionary, irrational and dangerous forces that have seized power in recent years? Is this the only way that we may work reasonably together against the states, groups and organizations that threaten our freedom and welfare?

Although current attacks on truth and justice give us reason to worry and speak out in rage, we are not convinced that they are best stopped by a disembodied return to detached scholarship and an abstract politics of formal rules and rights. Detached and disembodied research fails to understand the ugly desires and actions that drive and feed these attacks on our freedom and welfare; it fails to connect to the troubles and problems that people experience in everyday life; and it fails to understand how people may join forces without denying their bodily differences. As we will suggest in this chapter, there are other sensible responses to these problems, all of which imply that we must actively involve our bodies in the pursuit of truth and justice. This includes feminist responses, affective responses, Marxist responses and phenomenological responses.

EMBODIED STRUGGLE ACROSS FEMINIST AND AFFECTIVE POLITICS

From our previous discussions of feminist scholarship, it is no secret that women's bodies and embodied knowledge are a key starting point in feminist research and politics. Even though much feminist research lacks the explicitly embodied accounts which we have highlighted in the traditions of feminine writing (e.g. Cixous, 1976 [1975]; Irigaray, 1985 [1977]; Pullen, 2018) and Chicana feminism (e.g. Moraga and Anzaldúa, 1983), it is widely recognized across the feminist research community that the pursuit of women's rights is an embodied struggle which is informed by women's bodily experiences. As such, feminist politics is not simply an abstract question of equal opportunities, but a life-affirming battle against sexual violence, harassment and discrimination.

The women's movement of the 1960s and 1970s primarily sought liberation from marriage and other patriarchal structures that limited women's economic, political and sexual freedom (Delphy and Leonard, 1992). In contrast, the current #metoo movement has revealed horrific accounts that painfully alert us to the fact that women's equal opportunities depend on women's liberation from sexual violence, harassment and discrimination. The women who first spoke up in the #metoo movement shared dreadful experiences which struck a common chord and gave many others the courage to speak up. This shows that the personally embodied experience may be a source of knowledge, which mobilizes individual and collective affects with significant political effect.

Despite claims to the opposite, the affects and effects of body politics are not necessarily limited to particular identity groups.[1] A few months before the rise of the #metoo movement, the Women's March on 21 January 2017 filled the streets of numerous cities worldwide with thousands of people, and in some cities hundreds of thousands (*New York Times*, 2017; *USA Today*, 2017). In addition to fighting the backlash that US president Donald Trump was promising against women's rights, participants (many wearing self-knitted 'pussy hats') protested the reactionary attack that he was launching on human rights, racial equality, workers' rights, LGBT rights and the natural environment.

Some years prior to the Women's March and the rise of the #metoo movement, we witnessed another radical movement which thrived on a theory and practice of affective politics (especially inspired by the writings of Hardt and Negri, 2000, 2004): the Occupy movement. Joined by the slogan 'We are the 99%', Occupy involved 'a variety of different groups, including anti-consumerists, environmentalists, homeless activists, LGBT organizations, students, trade unionists, artists, hackers, ecological farmers, and small business coops' (Munro and Thanem, 2018a, p. 76).

Although the sharing of lived embodied experiences played a minor part in the Occupy movement's explicit agenda, many of its practices were profoundly embodied: when taking to the streets in spectacularly carnivalesque fashion to protest against greed, corruption, pollution and rising socioeconomic inequalities; when occupying central public squares in major cities around the world by setting up tents, kitchens and barricades; and when using hand signals rather than applause, shouts and other audible signals during meetings to avoid interrupting speakers.

PHYSICAL DEGRADATION, BODILY PAIN AND AFFECTIVE RESISTANCE IN ENGELS

The body has played a key role in a variety of political discourses. When reading the works of Karl Marx and Marxist theorists it may be close at hand to conclude that communist ideology and awakening rests on a philosophy and science of abstract concepts and technically complex analysis. Reading Friedrich Engels' main work, however, gives a rather different impression. In *The Conditions of the Working Class in England*, Engels (1958 [1845]) offers detailed descriptions of appalling physical conditions and misery.

Whereas Engels' accounts of mining and agricultural work are based on secondary sources (pamphlets, newspaper articles and the surgeon Peter Gaskell's book *The Manufacturing Population of England*), his discussion of working conditions in the textile industry are largely based on his first-hand observations from wandering the slums and factory areas of Manchester and Salford. For example, Engels offers the following account of needlewomen's (*sic*) work:

> The girls have to bend continually over their work and their food is both poor and difficult to digest. All this, and in particular the long hours of work and the lack of fresh air have the tragic result as far as the health of these girls is concerned. ... There they sit bent over their work and sew from four or five in the morning until midnight. Their health is ruined in a few years and they sink into an early grave, without having been able to earn the barest necessities of life. (pp. 237–8, 239)

The book contains a number of such accounts, all of which testify to the bodily effects of industrialization. Although Engels acknowledges how the extreme degradation of the working classes may cause passivity (see, e.g., p. 133), his rich descriptions feed directly into a larger and fundamentally affirmative political project.[2] Indeed, Engels implies that the lived experiences of the working classes may translate into powerful affects of solidarity and anger, which play a crucial role in revolutionary mobilization, protest and action. On the one hand, the fact that workers 'have experienced bitter suffering' themselves means that they 'are sympathetic to those in dire need' and 'more sociable and more friendly [towards those who are less fortunate than themselves]' (p. 140).[3] On the other hand, the inhuman treatment that they have been subjected to causes an anger and 'savage resentment ... against the upper classes' which 'shows that they are determined to free themselves without delay from the slavery imposed upon them' (pp. 132–3). Thus, Engels concludes towards the end of the book, 'There can be no doubt that the worker is well able to protest with all the fury at his [*sic*] command against the tyranny of the capitalist' (pp. 241–2).[4]

What we find in Engels, then, is very much an affective and embodied politics, which is expressed with vigour and intensity through the embodied troubles, experiences and struggles of other people.

PHILOSOPHICAL-POLITICAL ENGAGEMENT IN EMBODIED PHENOMENOLOGY

The connections between Marxist politics and lived embodiment are developed in a subtler but no less profound manner in the phenomenological and political writings of Maurice Merleau-Ponty. Merleau-Ponty, who had been a member of the French Resistance during the Second World War and introduced the pacifist Jean-Paul Sartre to Marxism, had admired the historical materialism of the late Marx for its capacity to reintegrate the economic subject and the living subject. When he later came to reject communism for its violence, he retained the scepticist openness of Marxism and continued its effort to write life into economy and history.

Just two years after Merleau-Ponty had published his major work *The Phenomenology of Perception* (1962 [1945]), he published his first political treatise, *Humanism and Terror* (1969 [1947]), which interrogates the terror that is committed by liberalism and communism in the name of humanism. While liberalism prides itself on pure principles of freedom and democracy, it typically turns a blind eye to the barbarism that follows in its footsteps, such as the colonial and postcolonial wars that are fought in pursuit of capitalist expansion and free enterprise.[5] In contrast, communism offers no illusions of being a non-violent ideology or movement, but assumes that 'establishing relations among men [*sic*] that are human' involves a 'revolutionary Stimmung of violence' (1969, p. xv).

Merleau-Ponty acknowledges that the consequences of liberalism need to be scrutinized but, more so, he questions if communism is 'still equal to its humanist intentions' (1969, p. xviii): the Moscow Trials of 1937 led him to argue that we can no longer just wait and see how communism will evolve;[6] a decade and a half later the Korean War convinced him that the USSR was an imperialist power without humanist intentions (Flynn, 2011).

However, even though Merleau-Ponty rejected communism and Sartre's one-sided defence of it,[7] he did not refuse Marxist theory as such. His foundational scepticism may actually have taken him closer to the questioning and open-minded attitude of Marxist philosophy than did the assertively inclined Sartre, who habitually spoke out in defence of communism (Goehr, 2005).[8] According to Coole (2007), Merleau-Ponty shows us that both communism and liberalism fail 'to understand the nature of the political or to recognize the particular virtuosity of the political actor' (p. 83). In contrast, 'Merleau-Ponty's work is ... political in the widest sense' (p. 83), as it seeks to transform our understanding of modern experience, away from the rationalist 'ethos of subjective mastery, ... reification and closure' (p. 82) which underpins liberalism and communism.

When Merleau-Ponty (1973 [1955]) later engages with Max Weber's liberal philosophy of history and with the teachings of Machiavelli, he finds a more flexible, critical and reasonable alternative to the historical determinism of Marxism and the economic reductionism of liberalism. These alternatives allow us to appreciate the virtuosity of political actors and recognize the complex, ambiguous and contingent nature of politics 'as it unfolds in the everyday' (p. 84). There is therefore a sincere effort in Merleau-Ponty's political writings to understand not just the nature of politics, but the nature of *life* in politics, and 'judge political regimes existentially ... according to the kind of lived relationships ... they ... support' (p. 86).

Thus, Merleau-Ponty's political writings bring forth the social and collective dimensions of his embodied phenomenology (Lefort, 2005). This is presaged in *The Phenomenology of Perception*, particularly when Merleau-Ponty emphasizes the feeling of affinity which underpins the revolutionary project.[9] As perception opens us up to the world, it throws us into a collective and political life where we live, act and struggle with others.

FROM BODY POLITICS TO AFFECTIVE REASON

The embodied phenomenology which runs through Merleau-Ponty's political writings and the physical accounts which flesh out Engels' main work launch a serious challenge against current appeals to disembody the pursuit of democracy and justice. Furthermore, the multiple alliances of the feminist and Occupy movements suggest that their embodied struggles by far transgress the selfish interests of individuals and identity groups. If we acknowledge that our bodies play a crucial role in advancing the democratic conditions of social and economic life, it makes no sense to conflate body politics with the narcissistic tendencies of identity politics.

However, at a time when the flourishing of fake news, alternative facts and rogue interpretations of statistical data has triggered new quests for 'real facts' and 'truth-telling', it is timely to consider what kinds of truth may be generated with embodied research. The widespread manipulation of news reporting, facts and statistical data gives no guarantee that quantitative methods will deliver untainted facts and objective truths, or that they will not be misinterpreted and misused by their users.

As we argued in Chapter 7, our analytical skills are deeply entangled with our embodied feelings and experiences. Seasoned statisticians and behavioural scientists would not dispute this. According to one experimental study where US research participants were asked to carefully consider both sides of the arguments on two hot political topics (gun control and affirmative action), polarization increased rather than decreased (Taber and Lodge, 2006, p. 760). Instead of considering both the pros and cons of the issues (and even though they had been 'instructed repeatedly to "set their feelings aside," to "rate the arguments fairly," and to be as "objective as possible"'), participants tended to 'mine' the readings, selecting arguments that supported their initial view whilst ignoring arguments that contradicted it.

In a neat piece on the role of feelings in statistics and reasoning, the economist and *Financial Times* journalist Tim Harford (2018) takes this to suggest that we need to be aware of our felt bias when processing new information. Hence, we should ask ourselves what we *feel* when being confronted with new information. Do we feel defensive, triumphant, angry, or perhaps relieved? 'If we don't at least acknowledge that we may be bringing some emotional baggage along with us', Harford argues, 'we have little chance of discerning what's true.' Furthermore, Harford maintains, we must stay curious, keep seeking information, use our judgement to evaluate how trustworthy the information is, and continue to ask open questions rather than get stuck in precise answers, which may prove to be wrong anyway.

Although these arguments acknowledge that no presentation of facts, information or knowledge can be purely objective or truthful, they imply a mind/body dichotomy which presumes that our embodied feelings and passions must be held in check as far as possible. In contrast, Merleau-Ponty (1962) maintained that no knowledge can be produced without the body (after all, the mind is in the body), and that any attempt to do so would cut us off from re-achieving direct contact with the world and gaining a phenomenologically truthful grasp of it.

While in Chapter 2 we noted how Merleau-Ponty's romantic view of lived embodiment leads him to reinforce the mind/body dualism which he first set out to debunk, accepting his premise that knowledge is embodied remains foundational in any effort to generate embodied truths about the everyday meandering of social and political life. A good place to continue this effort is to extend our discussion from Chapter 7, when we stressed the embodied nature of analytical thinking and rational decision-making, and to do so through the 17th-century philosophy of Benedict de Spinoza. This is less obscure than it may sound. The world-leading neuroscientist Antonio Damasio has already seen much value in Spinoza's work. But whereas Damasio (2003) primarily finds in Spinoza an insight that feelings are embodied in us because they help us make deliberate and life-saving choices (p. 175), it is worthwhile to remember that Spinoza can be pushed a notch or two further.

In his major work *The Ethics*, Spinoza (1994 [1677]) does not merely replace Descartes' mind/body dualism with a notion that the mind and body are parallel substances which are joined in the human being. He even emphasizes the embodied, social and political nature of reason (see also Thanem and Wallenberg, 2015). A central problem in Spinoza's *Ethics* concerns how to enhance our collective freedom. But to do so, we must first understand the *limits* of our freedom; that is, what causes us to act and what restricts our action. According to Spinoza, active affects such as love and reason enhance our freedom, and sad passions such as hatred and unreason limit it. Since we are dragged down by inadequate ideas and false images of thought which enslave us to sad passions, tradition and the whims and wills of others, Spinoza argues that we must cultivate our reason. It is only when we use our reason to develop adequate ideas that we may become the cause of our actions.[10]

For Spinoza, this is both an embodied process and a collective process. Acknowledging that our own powers are limited (p. 197), he argues that we can only enhance our freedom by forming friendships with others who strive to do the same (p. 217). Indeed, when led by reason we will recognize that our own welfare depends on the welfare of other people.[11] Furthermore, reason cannot be cultivated by trying to overcome or detach ourselves from our bodies. Since we can only know the causes of our actions through our bodies, Spinoza maintains, it is only through our embodied experiences that we may learn to pursue positive encounters with others, which enhance rather than limit our freedom (pp. 131, 138–9; see also Munro and Thanem, 2018b).[12] It is no easy task Spinoza leaves us with, but it is one worth pursuing.

SUMMARY: EMBODIED STRUGGLES, EMBODIED KNOWLEDGE

In this chapter, we have strived to show that there is no need to disembody the scholarly pursuit of knowledge and social justice. Embodied research methods are not in denial of reason, nor do they prevent us from exploring the political problems and challenges that are currently confronting us. Rather than being divided by our bodily differences, contemporary movements in feminist and affective politics provide ample evidence that the struggle for justice in no meagre sense is an embodied one: bodily experiences instigate political progress; political action involves physical gestures and movements; and people join forces to enhance our collective freedom without denying our bodily differences.

Whereas detached scholarship pursues an abstract notion of reason which fails to connect to the troubles and problems that people experience in everyday life, embodied research methods promote engaged scholarship which fleshes out the lived reality of social problems, and understands that our collective freedom and welfare rests on our capacity to cultivate our reason by learning from our embodied experiences.

Perhaps we need to work by way of the body particularly when confronting big political issues: the pains and sufferings of exploitation and oppression are felt by bodies, as are the pleasures and privileges of wealth and power; the effects of economic crisis are experienced as anxiety and exhaustion, while its causes are not unrelated to bodily desires of greed and lust; the droughts, floods and storms caused by climate change leave dead and diseased bodies behind, but are not isolated from the pleasures of meat eating and long-distance holidays and the everyday convenience offered by the automobile. While such forces affect different bodies asymmetrically, our pains and joys are not unique. As our own personal troubles resonate with larger social problems and public issues (Mills, 1959), what we feel in here is felt out there too.

Questions and exercises

1. Use a library or the internet to read up on a hot political topic that you are particularly passionate about. Dig out arguments that support your initial position and arguments that contradict it. What do you feel when reading supportive arguments? What do you feel when reading opposing arguments? Write down your feelings and visceral reactions. Then try to formulate a balanced account of the topic. Have a critical friend play devil's advocate and judge how well you have succeeded. How difficult, or easy, was it to detach yourself from your own passions?
2. Repeat the exercise, but now try and formulate an account of the topic which argues for the contrary position. Again, have a critical friend play devil's advocate and judge how well you have succeeded to promote the contrary position. Alternatively, do the exercise with a friend, colleague or fellow student, where both of you write a text supporting the contrary position, and then compare notes, playing devil's advocate against each other.
3. Having done both exercises, compare the two texts you have written. Which text is more persuasive? Which text is better written? Which text makes for the most interesting or enjoyable reading experience? Do any of the texts lead you to change or modify your own position? Which text does your critical friend prefer?

CONCLUSION

Embodied Imagination in Future Research

Research has always been embodied, and it always will be. Although the majority of social science research and methods texts have tended to pretend otherwise, the entire research process involves our bodies in some way, whether we are negotiating empirical access to groups and organizations, interviewing people in face-to-face meetings, immersing ourselves in the daily activities of our research participants, recording and analysing empirical findings, or writing and presenting our research.

To appreciate how profoundly embodied social life is and how deeply involved we are with our bodies in our attempts to study it, in this book we have engaged with a range of methodological practices, procedures and perspectives, all of which exploit the body's capacity to grasp and convey the rich nuances of lived experience and interaction in the meanderings of everyday life and work. In particular, we have sought to explore the bodily aspects and activities of ethnographic fieldwork and face-to-face interviews, examine the potential of visual and biographical methods to record and express bodily practices and experiences, and cut loose the analytical work we do from the disembodied presumptions that often contaminate or limit the sense of our empirical materials. Furthermore, we have engaged with embodied traditions of writing, discussed the physical aspects of the writing process, and considered how embodied knowledge may be presented and communicated to broader audiences without losing its bodily vicissitudes.

Throughout the book, we have discussed ethical dilemmas related to embodied research practices, especially the challenge of respecting research participants' integrity whilst exploring issues and experiences in their lives that are notably intimate and sensitive. However, ethics is not restricted to individual expressions of respect and responsibility. As Spinoza stressed in *The Ethics*, the task of leading a good, powerful and joyful life is a political problem which requires us to work together to enhance our common affects, freedom and welfare. Inevitably, this even concerns the research we do: how we are affected by our research participants, how we affect them, and how we affect people who read our research. That is why in this book we have sought to articulate an approach to research which explores and expresses the bodily nuances and intensities of social life through engaged accounts of the worlds and cultures we embody. By connecting the personal troubles of our research participants both to our own experiences and to social problems on a larger scale, it may become possible to craft vivid and powerful narratives that inspire political change to make different lives more liveable.

The embodied imagination and incarnation which this entails challenges us to engage in a meta-methodological discussion of what method is, what its limits are and what it can be. Above and beyond the practical matters of research methods as such, this concerns how we write about method and how we write about life and work in the social worlds we explore. Whereas the majority of methods texts follow a rather standard formula of frameworks, toolboxes, recipes and instructions, in this book we have sought to flesh out the dilemmas, challenges and frustrations that we experience when doing empirical research.

Furthermore, this requires us to think critically about what comes after method. In some sense, the embodied research methods we have sought to map out here incorporate an argument against method; they warn against fetishizing embodied methods in the same way that disembodied methods have been reified as guarantors of robust research and truthful knowledge. Indeed, the feminist, Marxist and other perspectives we have drawn on in this book remind us that method is not an end in itself but an opening to new movements in thought and practice. To take the embodied imagination seriously, we must not simply ask what a specific method does, what it makes us do and what it prevents us from doing. To find out what an embodied method can do, we must never stop asking what bodies can do.

NOTES

INTRODUCTION

1. Haynes (2012) found that women professionals in accounting and law who fail to present a professional demeanour (because they are regarded as overweight or because they are seen to dress too casually or in too trendy a manner) are subject to negative comments and sanctions by colleagues, managers and clients. In contrast, Brewis and Sinclair (2000) found that women social workers who embodied a butch identity were more effective at dealing with difficult clients.
2. Although some of the empirical research on trans people transitioning at work has emphasized the difficulty of actually making gender trouble (because cis colleagues often seek to help trans people pass as members of the opposite sex) (e.g. Schilt and Connell, 2007), Torkild's research with Louise Wallenberg suggests a different picture: whereas some cis colleagues were so troubled by trans colleagues who came out at work that they terminated long-term cordial relationships, others were so intrigued that they started reflecting about their own gender identity in new ways (Thanem and Wallenberg, 2016).
3. For instance, Sutton (2007) discusses how naked male activists at the World Social Forum in Porto Alegre in 2003 were chased by police on horses as they took to the streets to express solidarity for a young woman who had been jailed for allegedly bathing without clothes; Reestorff (2014) comments how members of the 'sextremist' feminist collective Femen use their young, feminine and apparently frail but tough topless bodies as posters in acts of culture jamming where they loudly and conspicuously protest the sexualization, exploitation and oppression of women's bodies; Puggioni's (2014) research on non-status migrants to Europe draws attention to the political potential and constraints of the body whose subjects lack opportunities to voice dissent, and resort to 'self-mutilation, hunger strike, suicide attempts, and lips and eyes sewing' when robbed of speech (p. 562).
4. In Marx's own words, the objectification of labour 'estranges from man his [*sic*] own body' (2009, p. 32).

CHAPTER 1

1. A similar inclination can be seen in Mike Featherstone's (1982) text 'The body in consumer culture', which focuses on the disciplinary mechanisms through which consumers seek to take care of and improve their bodies.

2. Somewhat ironically, in his book *Regulating Bodies: Essays in Medical Sociology*, Turner (1992) makes further efforts to nuance his concept of the body by highlighting the body's transgressive and experiential nature; for example, he claims profound inspiration from Nietzsche's notion that ' "Man" is an unfinished product' (p. 2) as well as Nietzsche's contrast between the rational Apollo and the irrational Dionysus (p. 21). However, the book largely focuses on the rational and regulatory practices of medicine and how these are actualized in dietary management, the anatomy lesson and the eating disorder anorexia nervosa.
3. Other examples include Lloyd's (1996) study of how aerobics is constituted as a dominant fitness regime for western women through populist and expert advice communicated in glossy magazines and books, and Petersen's (1998) study of the standard, multi-edition anatomy textbook *Gray's Anatomy*. Examining *Gray's Anatomy*, from its first edition of 1858 to what was then the most recent edition of 1995, Petersen shows how the male and female sexual organs and the sexed body more broadly come to be textually and visually constructed as binary opposites.
4. As Stoller (1997) argues, Foucault (1979, 1980) and Butler (1990) deconstruct the Cartesian binary between mind and body, but 'their bloodless language reinforces the very principle they critique ... [and thus] ... regulates and subjugates the very bodies they would liberate' (p. xv). If this is so for authors who radically refuse the normative legacy of disciplines and discourse, it is much worse in the social sciences more generally.
5. To readers in ancient times the *Phaedo* was known under the title *On the Soul*. The dialogue is claimed to record the last day of Socrates' life, as he was awaiting trial by the city of Athens, accused of impiety and corrupting the youth.
6. When referencing the works of classical philosophers such as Plato, and later Aristotle, we reference books and line numbers rather than page numbers because their works are available in a large number of different print editions as well as from a number of different internet sites. Thus, we are following the standard style of referencing among specialists in philosophy. Quotes from Plato's *Phaedo* are from Hackforth's (1955) translation. Quotes from Plato's *The Republic* are from Lee's (1987) translation.
7. Plato's negative view of the body is even more strongly expressed as he calls the body an 'evil admixture' which confuses us, disturbs us and alarms us, 'makes countless demands upon us', 'befouls' our souls and causes us to go to war to 'acquire riches' and satisfy its desires, 'hampers our pursuit of true being', and 'fills us with ... such quantities of trash, that ... we really never have a moment to think about anything' (*Phaedo*, 66b–d).
8. Plato argues that it is only at death, when we are parted from the body and only the immortal soul remains, that we can attain true knowledge, because it is only when we have 'escape[d] from the body' that we may 'observe things in themselves with the soul by itself' (*Phaedo*, 66e).
9. Although the Plato commentator Jerome Neu (1971) has critiqued previous claims (perhaps most famously by Popper, 1962) that *The Republic* offers an organic theory of the state, there are hints of organicism when Plato discusses the importance of, and grounds for, social cohesion in Book V. Here, he argues that 'the best-ordered state ... most nearly resembles an individual', because 'when one of us hurts his finger ... the whole organism ... perceives it' (462c); 'such a community will regard the individual who experiences gain or loss as part of itself, and be glad or sorry as a whole accordingly' (462e).

10. According to the close reading conducted by Gail Fine (1993), *On Ideas* is the only Aristotelian text which includes a systematic critique of Plato's theory of forms, but there is lacking consensus amongst Aristotelian commentators that this text was actually written by Aristotle as it only survives in fragments penned by Alexander of Aphrodisias.
11. In *On the Soul* Aristotle defines substance as 'form, matter, and the complex of both', adding that 'of these matter is potentiality, form actuality' (414a4–414a28). These ideas gain biological application in *On the Generation of Animals*, where Aristotle develops a 'hylomorphic semenology' which renders life a matter of substance shaped and realized according to a predestined plan. Here, Aristotle even posits a hideous theory of human reproduction where life is wholly embedded within the semen and where the womb is reduced to an empty container wherein the semen comes to full fruition (767b15–17; see also Protevi, 2001, p. 79).
12. 'For the soul rules the body with the rule of the master, whereas understanding [i.e. reason] rules desire with the rule of a statesman ... it is evident that [this] is natural and beneficial' (1254b5); meanwhile, 'the rule of a master is not the same as rule of a statesman ... For rule of a statesman is rule over people who are naturally free, whereas that of a master is rule over slaves' (1255b15).
13. The *cogito* is reiterated in Part 1 (section 7) of Descartes' *Principles of Philosophy*, where he sets out 'The principles of human knowledge'. Although some commentators have emphasized that Descartes in the *Meditations* merely states that 'I think, I exist' (*'je pense, je suis'*) (Descartes 1996, p. 17; see, e.g., Newman, 2016), this does not seem to have been fully acknowledged across the social sciences. Rather, the profound connection between consciousness, subjectivity and existence that has been implied by the above translation seems to have prevailed throughout western intellectual history.
14. According to certain commentators, such as the medical philosopher Stuart Spicker (1970), Descartes was not himself a Cartesian, at least not insofar as that implies believing in the mind's ability to control the body and the distinct nature of the mind and the body. When responding to the philosophical interrogations lettered by Princess Elizabeth of Bohemia, Descartes ended up contradicting his own separation of mind and body by admitting that the mind–body union can be understood through the senses and that the body can act on the mind through bodily awareness. Descartes acknowledged this, albeit abruptly, in Meditation Six, Spicker argues, when stating that the body may act on the mind through feelings of pain. In a letter to his follower Regius, Descartes goes even further, claiming that 'It is not accidental to the human body to be united to the mind but its very nature' (cited in Spicker, 1970, p. 17).
15. In *Treatise on Man*, Descartes (1633) emphasized that several bodily functions in human beings, such as digestion, heart beat and perception, involved no intervention from the mind, and in *Description of the Human Body* Descartes (1647) rejected any claims that the soul could move the body because the body was a mechanical entity which moved as a result of the heat of the heart.
16. Whereas Copernicus' original turn from geocentric to heliocentric astronomy meant that the earth revolved around the sun rather than *vice versa*, Kant (1998 [1787]) articulated, in the preface to the second edition of the *Critique of Pure Reason* (Bxvi), an anthropocentric turn which made the world revolve around the human.

17. In the 'Transcendental Dialectic' it also becomes apparent that Kant's notion of the transcendental ego, which synthesizes our sensations according to the categories of reason, is not a substantial 'thing' like the soul in Descartes, but an ideal condition.
18. It has even been argued that the soul–body relationship was Kant's overarching concern in his pre-Copernican (i.e. pre-1770) writings (Laywine, 1993). As Palmquist (2016) shows in his reading of *Dreams of a Spirit-Seer*, Kant, when discussing if the soul can be located in a specific part of the body (pp. 324–5), argues that he feels himself as much present in his fingertips and his aching heel as in his heart and head (see Palmquist, 2016, p. 200). A similar concern may be seen in Kant's Copernican writings. In their introduction to the standard edition of Kant's *Critique of Pure Reason*, Guyer and Wood (1998, p. 72) emphasize that Kant, *contra* Descartes, holds 'that there should be no puzzle about the possibility of interaction between mind and body because the differences in their appearances that Descartes and his followers had assumed to stand in the way of interaction might be no more than different appearances of a single sort of underlying reality'.
19. Lukes (1982, p. 11) actually suggests that Durkheim's (1982 [1895]) *Rules of Sociological Method* echoes Descartes' *Discourse on Method*.
20. Here, Durkheim compares his method for the observation of social facts with 'Descartes' method of doubt', which he views as 'only an application of' the rule that '*One must systematically discard all preconceptions*' (1982, p. 72; original emphasis).
21. This is reasserted in *Suicide*, where Durkheim states that 'For sociology to be possible it must ... have an object all of its own' (1951 [1897], p. 59).
22. Elaborating this point, Durkheim explained that 'The pressure exerted by one or several bodies on other bodies ... should not be confused with that which group consciousness exercises on the consciousness of its members.' Whereas habits are within us, 'social beliefs and practices act upon us from the outside' (1982, p. 44).
23. In fact, Durkheim's (1951 [1897]) study of suicide is a good example of how he engaged with mixed phenomena in this way, explaining increasing suicide rates in modern society with increasing *anomie* and lacking moral guidance.
24. This epistemic and anthropological dualism was preceded by a social dualism in Durkheim's (1984 [1893]) doctoral thesis, *The Division of Labour in Society*, where he argued that such division of labour was produced through mechanical and organic solidarity.

CHAPTER 2

1. These are the examples Merleau-Ponty offers on p. 167 of the *Phenomenology*.
2. Recollecting Descartes' correspondence with Princess Elisabeth of Bohemia (p. 230), Merleau-Ponty argues that even though 'I am my body ... to the extent that I possess experience', the fact that my body is also a '"natural" subject' (p. 231) means that my experience of my body 'runs counter to the reflective procedure which detaches subject and object from each other, and which gives us only the thought about the body, or the body as idea and not the experience of the body or the body in reality'.
3. For Merleau-Ponty (1962) this is related to the sense in which primary perception is a 'non-thetic', pre-objective and pre-conscious experience (p. 281).

4. Perhaps this should not come as a surprise. Although Merleau-Ponty is deeply critical of what he regards as Kant's Copernican reduction of the world in itself to our phenomenal knowledge of the world, he is a phenomenologist fully concerned with the limits of human knowledge.
5. Shusterman's (2005) examples of this include visualizing what some part of the body feels like, or visualizing how it moves and how it is oriented (p. 156).
6. Thus, 'somatic awareness' is rather different from the Merleau-Pontian concept of the increasingly internalized, habituated and unconscious 'skilful coping', which was developed by Hubert and Stuart Dreyfus (1982, 2005) and which has become increasingly popular among many of our colleagues keen to understand the embodied dimension of knowledge generation in organizations (see, e.g., Hindmarsh and Pilnick, 2007; Yakhlef, 2010).
7. Further examples of disability research engaging with Merleau-Ponty include Paterson and Hughes (1999), Toombs (2001), Hughes (2007) and Cadwallader (2010). Feminist engagements with his work include Heinämaa (2003), Olkowski and Weiss (2006) and Hamington (2008). Studies of ageing using his work include Rowles (1978) and Schwanen et al. (2012). For a recent discussion of racism using his work readers may want to take a look at Ngo (2016).
8. The situation is different in literary fiction, which includes several accounts of the real embodied lives of people suffering from deprivation. For instance, Robert Tressell's (1914) novel *The Ragged-Trousered Philanthropists* captures much of the poverty, physical hardship and emaciation of working-class life at the turn of the 19th century. So does George Orwell's (1933) *Down and Out in Paris and London.*
9. Merleau-Ponty (1962) calls this the 'eidetic reduction' of phenomenology, which involves identifying the necessary essence of a phenomenon, that is, looking for the world's essence as a fact for us before any thematization rather than looking for the world's essence as a universal idea (pp. xvi–xvii).

CHAPTER 4

1. This department employed sales representatives whose task was primarily to fill the newspaper with what is called run-of-paper advertising. Such advertising appears within the editorial pages in contrast to classified advertising that appears at the back of the paper, and is sold through a call centre but where the customer ordinarily approaches the newspaper rather than the other way around. Research was conducted in this call centre by my PhD student John Roberts some of which was published in Knights and Roberts (1982).
2. This was a time well before I had become interested in questions of masculinity so like most men was unreflectively ensconced in it. Later academic and then personal biographical conditions led me to challenge its oppressive and repressive effects.
3. A part of this was published in Knights and Roberts (1982).
4. Czarniawska actually identifies an even earlier use of shadowing in a study by Guest (1955), where he 'observe[d] a foreman for a full eight-hour day and ... record[ed] everything he did' (in Czarniawska, 2007, p. 22). Like Capote, Guest never used the term shadowing.

5. Actually, this CEO resembled the one in the sportswear company studied by Torkild. He was immaculately dressed, played squash competitively and made it clear to all his managers that they should aim to emulate him, and he was aggressively demanding and denigrated any weakness in a manager in public. Interestingly, the company collapsed a few years after the research ended, suggesting that it can be counterproductive for aggressively competitive managers to impose their values on the organization.

CHAPTER 5

1. For example, Darwin suggests that confusion can spur different bodily responses in different people: whereas 'A vulgar man often scratches his head when perplexed in mind ... Another man runs his eyes ...' (1872, p. 32).

CHAPTER 6

1. Even in contemporary culture, we tend not to eat in front of those who are not eating. As David admits: 'I for one feel terribly uncomfortable and embarrassed if someone enters the room where I'm eating. Whether or not this is connected to the Balinese culture of shame I'm unsure but maybe it's related to the idea that if there is food being consumed we should share it with anyone present. However, if this is a casual visitor or a trades person doing some work in the house, it may not be functional or socially appropriate to do so and what I would ordinarily do is to put the food away during the brief encounter with the person. Certainly, I'd be offended if the person instead of apologizing for the interruption began photographing me.'
2. In my own case, the heavy alcohol abuse was made slightly more tolerable as the leader of this union group enjoyed talking with me as he had ambitions to apply to study for a university degree. At his request, after 16 months of engaged observations, naïvely I wrote a report for the staff that outlined how I thought they could escape from the impasse but this resulted in my expulsion since they feared that I would pass on my report to management who would then have direct evidence to execute a redundancy programme that had been earlier aborted because of union intervention. I say, 'naïvely' because I felt that I had not completed my research. Hence, I would have preferred to remain a little longer, although my professor argued that it was the best outcome as I might never have thought it was an appropriate time to leave the field. My lesson was that subsequently I never fed back reports until having voluntarily left the field.
3. Such diary records are in stark contrast to professional diaries and calendars, such as those collected by the management scholar Sune Carlson (1991 [1951]) when studying executive behaviour. While Carlson's unique access to the diaries of Swedish executives yielded significant insights into their daily routines and activities (including telephone calls, meetings and the kinds of questions handled), the diaries as such (as well as Carlson's engagement with them) largely consisted of short, impersonal

and disembodied annotations, all of which were not even kept by the executives themselves but by secretaries and personal assistants.

4. In recent years, there have even been attempts to combine biographical and visual methods to generate more explicitly sensuous and embodied accounts from research participants. For instance, scholars in health and illness studies have used video diaries in ways that enable participants to both 'show and tell' about their lived experience of a certain health condition, and enable researchers to see, hear and feel the research participant's body within one media space (see, e.g., Bates, 2013, in Bartlett and Milligan, 2015).
5. On 12 November 1668 Pepys made the following entry: 'And so having dined ... I to my wife and to sit with her a little; and then called her and Willet to my chamber, and there did with tears in my eyes, which I could not help, discharge her and advise her to be gone as soon as she could, and never to see me or let me see her more while she was in the house; which she took with tears too' (2003, p. 960).
6. Torkild was certainly overwhelmed with guilty laughter when reading her portrait of 'Maynard', as sporting 'a queer swollen eel-like look, not very pleasant' and as having 'grown very gross and stout, especially when he wraps his leopard-spotted dressing-gown tightly around his knees and stands in front of the fire' (11 September 1923, p. 168).
7. Yet conventional diaries are not the only vehicle for recording or perhaps imaginatively extemporizing one's experiences. As we were writing this, David had just read one of the greatest literary examples of this: Marcel Proust's (2016 [1913–27]) *Remembrance of Things Past* where some of the most vivid pictures of youthful love are elaborated.

CHAPTER 7

1. For instance, visiting the website of the Boston Consulting Group, we learn that their Analytics Team consists of 'data engineers, data scientists, app developers, and technical experts' who are professionals in 'acquiring client data, cleaning and summarizing it for easy analysis, ... big data management, modeling and tool support, and developing custom client tools and applications ... This team also trains consulting staff on advanced analytics tools including Alteryx, Tableau, and Meridian as part of our Analytics University' (www.bcg.com/careers/path/knowledge-analytics/roles.aspx).
2. In his early dialogues, Plato has Socrates ask a variety of questions to find out what virtue is, cross-examining his conversation partners. As Beaney (2017, n.p.) reminds us, in the *Laches* Plato has Socrates ask 'What is courage [*andreia*]?', and in the *Euthyphro* he has him ask 'What is piety [*hosiotês*]?'
3. Indeed, *Meno*, and Meno's paradox, is the turning point in Plato's analytical method. Having Socrates ask 'What is virtue [*aretê*]?', Plato realizes that either you know something or you don't, but in both cases the search for the truth stops: if you know something there is nothing more to search for, and if you do not, you do not even know what to search for.
4. In modern terms, one might say that sub-categories can only be defined by cutting off features from overarching categories.

5. In her book, which originally was titled 'Under the Knife', Dale (2001) demonstrates how the academic field of organization studies has been founded on an anatomical logic of dissection.
6. In the *Treatise on Man* (1633) and the *Discourse on Method* (1637) Descartes avers that the mechanistic nature of the human organism paralleled that of animals, the entire universe, and that of contemporary inventions such as clockworks, mechanical pumps and water fountains.
7. Other famous analysts have involved the body in equally surgical but more perverse ways. Freudian psychoanalysis involves a number of bodily analogies, assumptions and requirements. Not only does psychoanalysis seek to access the unconscious during lengthy sessions where the patient is required to lie down on a couch without facing the psychoanalyst, or trace psychological development to childhood sexuality which is categorized according to the bodily analogies of the oral, the anal and the genital phase. At one stage, Freud even believed that sexual problems were connected to the anatomy of the nose, a view which was promoted by his friend and collaborator the surgeon Wilhelm Fliess. When seeking to treat hysteria in Freud's patient Emma Eckstein (a hysteria which Freud explained was caused by excessive masturbation), Fliess conducted a surgical procedure (with local anaesthetic) on Eckstein's nasal cavity which left her permanently disfigured. Although Freud was embarrassed by the incident, he later proclaimed confidence in Fliess' competence as a medical professional, and blamed Eckstein for her post-operative haemorrhages, arguing that they were 'wish-bleedings' she had caused in pursuit of the affection of others (Gay, 2006 [1989], pp. 84–7, 154–6).
8. Damasio takes inspiration from the historical case of the New England railway worker Phineas Gage, who in 1848 survived an accident were an iron rod went through his head and damaged his left frontal lobe and his capacity to feel pain and pleasure.
9. Although this has led commentators (e.g. in popular outlets such as *Psychology Today*) to argue that being analytical requires us to be intuitive, the division between analytical and intuitive thinking is reinforced by the behavioural scientist and Nobel Laureate Daniel Kahnemann. In *Thinking Fast and Slow* (2011), which summarizes his life's work, Kahnemann encourages readers to abandon our tendency for fast, intuitive thinking and gut-based decision-making, and instead learn to think slowly, logically, analytically and rationally, by cultivating our ability to stay calm and keep our cool. On this view, having 'an analytical mind' typically implies that one has the ability to identify clear patterns and causal relationships by thinking calmly and logically about a problem or a phenomenon rather than relying on one's intuition and initial 'gut feeling'.
10. When we were trying to analyse our empirical material to understand how transvestites do and undo gender (see Thanem and Wallenberg, 2016), we found existing concepts of gender and transgender limited. Comparing and contrasting our material with previous research in the area, we noticed that our research participants neither copied feminine stereotypes in order to pass as women, nor did they exaggerate such stereotype through hyperbolic drag acts. While we started the process heavily inspired by Judith Butler's (1990, 1993, 2004) work on gender as performance, our findings did not quite fit her notion that gender is done and undone through binary and excessive forms of drag. Most of our participants had insisted that they did not crossdress in order to pass as

members of the opposite sex, but they had also distanced themselves from drag. They tended to dress casually, and they did not change their voice or body language when crossdressing. Indeed, transvestites seemed to express transgender by *underdoing* gender rather than simply doing or overdoing gender: by wearing feminine clothes, hairstyles and make-up whilst retaining a masculine voice and daring to do so by cultivating a kind of self-confidence which we were unable to squeeze into the gender binary.

11. More specifically, Rich emphasizes that her feminist analysis is 'tangled with parts of my life which stayed buried even while I dug away at the strata of early childhood, adolescence, separation from parents, my vocation as a poet; the geographies of marriage, spiritual divorce, and death, through which I entered the open ground of middle age' (1976, p. 15).
12. More recently, Sara Ahmed (2006, p. 10) points out in the introduction to her queer phenomenological engagement with migrant orientation that 'it is my own experience as a migrant subject ... that has led me to think about orientation and to wonder about how it is that we come to inhabit spaces as if they extend our skin'.
13. Interestingly, the etymology further embodies the practice of textual interpretation. As a text has texture, it is a material and corporeal being with maternal roots – woven into a matrix by an umbilical cord spun out of our mother's uterus (see Dale, 2001).

CHAPTER 8

1. Mind you, as Dennis Tourish (2015) has shown, there are still plenty of examples of researchers dressing up unoriginal findings and self-evident facts in unreadable jargon, and getting them published in some of the most prestigious journals.
2. David interjects: 'Giddens may be a prolific writer and a brilliant theorist but less of an empirical researcher. I remember one time we were walking in the streets of Dubrovnik after Giddens had just given a superb lecture to postgraduate students. Then he indicates that he ought to do some empirical research, and he asks me for advice as to how to go about it!'

CHAPTER 9

1. Cixous (1976) for one emphasizes that there is no contradiction between feminist struggle and class struggle: 'Not that in order to be a woman-in-struggle(s) you have to leave the class struggle or repudiate it; but you have to split it open, spread it out, push it forward, fill it with the fundamental struggle so as to prevent the class struggle, or any other struggle for the liberation of a class or people, from operating as a form of repression, pretext for postponing the inevitable, the staggering alteration in power relations and in the production of individualities' (pp. 882–3).
2. 'If people are relegated to the position of animals', Engels maintains, 'they are left with the alternatives of revolting or sinking into bestiality' (1958, p. 144).

3. The version available online at marxists.org articulates this in an even more embodied tone: 'the humanity of the workers is constantly manifesting itself pleasantly. They have experienced hard times themselves, and can therefore feel for those in trouble, whence they are more approachable, friendlier, and less greedy for money, though they need it far more than the property-holding class.'
4. Here, Engels also notes that whereas the bourgeoisie's 'ruling passions' of selfishness, greed and 'love of money' have been cultivated through their emphasis on 'the reasoning faculty', the worker's passions, 'Given free rein', will 'readily stimulate him [*sic*] to action' (pp. 241–2).
5. Behind the 'mystification of liberalism', Merleau-Ponty argues, 'The purity of principles not only tolerates but even requires violence' (1969, p. xiii).
6. Like Trotsky admitted, we need to give communism a deadline (Lefort, 2005, p. 374).
7. Merleau-Ponty offered his silent response against Sartre's defence of communism when declaring, in the 1952 July issue, that he was leaving the journal *Les Temps Modernes*, which they had co-founded in 1945. His articulated response came in the essay 'Sartre and Ultrabolshevism', which was published in his second collection of political writings *Adventures of the Dialectic* (1973 [1955]).
8. Differentiating between Merleau-Ponty's philosophical-political engagement and Sartre's political commitment, Goehr (2005) suggests that Merleau-Ponty's political engagement even comes through in his turn to silence during the Korean War; where politics is reduced to violence, Merleau-Ponty (1952) argues that philosophy must withdraw into silence.
9. Merleau-Ponty illustrates the feeling of affinity through the example of the Moscow 1917 peasant revolution, and argues (against rationalist objectivism and idealist subjectivism) that 'It is not necessary to think' but 'sufficient to feel a shared project' (p. 517). He also argues that class consciousness is neither the result of 'mechanical causality, of my status as workman or bourgeois' nor 'an unwarranted evaluation', but 'prepared by some molecular process' through which 'it matures in co-existence before bursting forth into words and being related to objective ends' (p. 446). In both examples, he acknowledges the role that living labour played in changing the conditions of social and economic life.
10. Spinoza notes that 'Man's [*sic*] true power of acting, or virtue, is reason itself ... I call him free who is led by reason alone' (1994, p. 235).
11. Spinoza terms this the 'principle of common life and common advantage' whereby 'everyone who is led by reason desires for others also the good he wants for himself' (1994, p. 238).
12. Thus, seeking to overcome the appetites of our bodies would not only be impossible, but futile and self-defeating because it would take us further away from understanding what causes us to act, and from understanding how we might pursue joyful encounters with others that enhance what we can do.

REFERENCES

Acker, J. (1990) 'Hierarchies, jobs and bodies: A theory of gendered organizations', *Gender & Society*, *4*(2), 139–58.

Ahmed, S. (2004) *The Cultural Politics of Emotions*. London: Routledge.

Ahmed, S. (2006) *Queer Phenomenology: Orientations, Objects, Others*. Durham, NC: Duke University Press.

Ahmed, S. (2007) 'A phenomenology of whiteness', *Feminist Theory*, *8*(2), 149–68.

Ahmed, S. (2017) *Living a Feminist Life*. Durham, NC: Duke University Press.

Allen, C. (2004) 'Merleau-Ponty's phenomenology and the body-in-space encounters of visually impaired children', *Environment and Planning D: Society and Space*, *22*(5), 719–35.

Allen-Collinson, J. (2011) 'Intention and epoche in tension: Autophenomeno-graphy, bracketing and a novel approach to researching sporting embodiment', *Qualitative Research in Sport, Exercise & Health*, *3*(1), 48–62.

Allen-Collinson, J. (2013) 'Narratives of and from a running-woman's body: Feminist phenomenological perspectives on running embodiment', *Leisure Studies Association Newsletter*, *95*, 41–8.

Allport, G. (1942) *The Use of Personal Documents in Psychological Science*. New York: Social Science Research Council.

Alvesson, M. and Sköldberg, K. (2009) *Reflexive Methodology*, 2nd edition. London: Sage.

Ambjörnsson, F. (2004) *I en klass för sig: Genus, klass och sexualitet bland gymnasietjejer*. Stockholm: Ordfront.

Angelou, M. (1990) 'Maya Angelou, The art of fiction No. 119', interviewed by G. Plimpton, *The Paris Review*, Issue 116 (Fall). www.theparisreview.org/interviews/2279/maya-angelou-the-art-of-fiction-no-119-maya-angelou

Antoninetti, M. and Garrett, M. (2012) 'Body capital and the geography of aging', *Area*, *44*(3), 364–70.

Aristotle (1991) *The Complete Works of Aristotle*. The revised Oxford translation, edited by J. Barnes. Princeton, NJ: Princeton University Press.

Aristotle (1998) *Politics*, translated by C. D. C. Reeve. Indianapolis, IN: Hackett.

Atkinson, P., Coffey, A., Delamont, S., Lofland, J. and Lofland, L. (2001) *Handbook of Ethnography*. London: Sage.
Atwood, M. (2013) 'How I write: Margaret Atwood', interviewed by N. Charney, *The Daily Beast*, 10 October. www.thedailybeast.com/how-i-write-margaret-atwood
Aull Davies, C. (2008) *Reflexive Ethnography: A Guide to Researching Selves and Others*, 2nd edition. Abingdon: Routledge.
Bacon, F. (1858 [1620]) *The Works of Francis Bacon, Vol. IV*, collected and edited by J. Spedding, R. L. Ellis and D. D. Heath. London: Longman & Co. et al.
Banks, M. (2001) *Visual Methods in Social Research*. London: Sage.
Bartlett, R. and Milligan, C. (2015) *What is Diary Method?* London: Bloomsbury.
Bateson, G. and Mead, M. (1942) *Balinese Character: A Photographic Analysis*. New York: New York Academy of Sciences.
Battersby, C. (1989) *Gender and Genius: Towards a Feminist Aesthetics*. London: Women's Press.
Beaney, M. (2017) 'Analysis', in E. N. Zalta (ed.) *The Stanford Encyclopedia of Philosophy*, winter 2017 edition. https://plato.stanford.edu/archives/win2017/entries/analysis/
Beattie, G. (2003) *Visible Thought: The New Psychology of Body Language*. London: Routledge.
Beavan, K. (2017) 'Sleepless and inchoate in Boston: A performative auto/ethnography', presentation at the 33rd EGOS Conference, 6–8 July, Copenhagen, Denmark.
Becker, H. S. (1953) 'Becoming a marihuana user', *American Journal of Sociology*, *59*(3), 235–42.
Becker, H. S. (1973) *Outsiders: Studies in the Sociology of Deviance*. New York: The Free Press.
Becker, H. S. (1974) 'Photography and sociology', *Studies in Visual Communication*, *1*(1), 3–26.
Becker, H. S. (1998) *Tricks of the Trade: How to Think About Your Research While You're Doing It*. Chicago: University of Chicago Press.
Becker, H. S. (2002) 'Visual evidence: *A Seventh Man*, the specified generalization, and the work of the reader', *Visual Studies*, *17*(1), 3–11.
Becker, H. S. (2007 [1986]) *Writing for Social Scientists*. Chicago: University of Chicago Press.
Behar, R. (1996) *The Vulnerable Observer: Anthropology That Breaks Your Heart*. Boston: Beacon Press.
Bell, C. (1844) *The Anatomy and Philosophy of Expression as Connected with the Fine Arts*. London: John Murray.
Bell, E. and King, D. (2010) 'The elephant in the room: Critical Management Studies conferences as a site of body pedagogics', *Management Learning*, *41*(4), 429–42.
Berggren, N., Jordahl, H. and Poutvaara, P. (2010) 'The looks of a winner: Beauty and electoral success', *Journal of Public Economics*, *94*(1), 8–15.
Bigo, V. and Islam, G. (2018) 'Practices of inspiration: Leadership, yoga, and the organization of embodiment', unpublished paper, Kedge Business School, Marseille, France.

Bolger, N., Davis, A. and Rafaeli, E. (2003) 'Diary methods: Capturing life as it is lived', *Annual Review of Psychology*, *54*(1), 579–616.

Bordo, S. (1993) *Unbearable Weight: Feminism, Western Culture, and the Body*. Berkeley, CA: University of California Press.

Bourdieu, P. (1977) *Outline of a Theory of Practice*. Cambridge: Cambridge University Press.

Bourdieu, P. (1984 [1979]) *Distinction: A Social Critique of the Judgment of Taste*. Cambridge, MA: Harvard University Press.

Braverman, H. (1973) *Labor and Monopoly Capital: The Degradation of Work in the Twentieth Century*. New York: Monthly Review Press.

Brewis, J. and Sinclair, J. (2000) 'Exploring embodiment: Women, biology and work', in J. Hassard, R. Holliday and H. Willmott (eds) *Body and Organization*, pp. 192–214. London: Sage.

Brewis, J., Meisenbach, R., Rippin, A., Risberg, A., Sayers, J. and Sköld, D. (eds) (2017) 'Professor Heather Höpfl, 1948–2014: Eine Gedenkschrift', *Culture & Organization*, *22*(2), 81–156.

Broom, A. F., Kirby, E. R., Adams, J. and Refshauge, K. M. (2015) 'On illegitimacy, suffering and recognition: A diary study of women living with chronic pain', *Sociology*, *49*(4), 712–31.

Bryman, A. (2012) *Social Research Methods*, 4th edition. Oxford: Oxford University Press.

Burke, S. M., Sparkes, A. C. and Allen-Collinson, J. (2008) 'High altitude climbers as ethnomethodologists making sense of cognitive dissonance: Ethnographic insights from an attempt to scale Mt. Everest', *The Sport Psychologist*, *22*, 336–55.

Butler, J. (1990) *Gender Trouble: Feminism and the Subversion of Identity*. London: Routledge.

Butler, J. (1993) *Bodies that Matter: On the Discursive Limits of 'Sex'*. New York: Routledge.

Butler, J. (2004) *Undoing Gender*. New York: Routledge.

Byrne, P. H. (1997) *Analysis and Science in Aristotle*. Albany: SUNY Press.

Cadwallader, J. R. (2010) 'Stirring up the sediment: The corporeal pedagogies of disabilities', *Discourse: Studies in the Cultural Politics of Education*, *31*(4), 513–26.

Capote, T. (1980) 'A day's work', in *Music for Chameleons*. New York: Random House.

Carlson, S. (1991 [1951]) *Executive Behaviour*. Uppsala: Acta Universitatis Upsaliensis (Studia Oeconomiae Negotiorum).

Cederström, C. and Spicer, A. (2017) *Desperately Seeking Self-Improvement: A Year Inside the Optimization Movement*. London: OR Books.

Cederström, C. and Thanem, T. (2018) 'The Swedish CEO who runs his company like a CrossFit gym', *Harvard Business Review*, 12 March. https://hbr.org/2018/03/the-swedish-ceo-who-runs-his-company-like-a-crossfit-gym

Charlesworth, S. J. (1999) *A Phenomenology of Working Class Experience*. Cambridge: Cambridge University Press.

Chia, R. and Kallinikos, J. (1998) 'Epilogue: Interview with Robert Cooper', in R. Chia (ed.) *Organized Worlds: Explorations in Technology and Organization with Robert Cooper*, pp. 131–80. London: Routledge.

Cixous, H. (1976 [1975]) 'The laugh of Medusa', translated by K. Cohen and P. Cohen, *Signs*, *1*(4), 875–93.

Clarke, C. and Knights, D. (2015) 'Careering through academia: Identity and ethical subjectivity in UK business schools', *Human Relations*, *68*(12), 1865–88.

Clarke, C. and Knights, D. (2018) 'Practice makes perfect? Skillful performance in veterinary work', *Human Relations*, *71*(10), 1395–421.

Clifford, J. and Marcus, G. E. (eds) (1986) *Writing Culture: The Poetics and Politics of Ethnography*. Berkeley, CA: University of California Press.

Cohen, D., Beattie, G. and Shovelton, H. (2010) 'Nonverbal indicators of deception: How iconic gestures reveal thoughts that cannot be suppressed', *Semiotica*, *182*, 133–74.

Collier, J. (1957) 'Photography in anthropology: A report on two experiments', *American Anthropologist*, *59*(5), 843–59.

Collier, J. and Collier, M. (1986) *Visual Anthropology: Photography as a Research Method*. Albuquerque, NM: University of New Mexico Press.

Coole, D. (2007) *Merleau-Ponty and Modern Politics After Anti-Humanism*. Lanham, MD: Rowman & Littlefield.

Cooper, C. and Dewe, P. (2008) 'Well-being – absenteeism, presenteeism, costs and challenges', *Occupational Medicine*, *58*(8), 522–4.

Cooper, R. (1986) 'Organization/disorganization', *Social Science Information*, *25*, 299–335.

Corballis, M. C. (2002) *From Hand to Mouth: The Origins of Language*. Princeton, NJ: Princeton University Press.

Cordaro, D. T., Sun, R., Keltner, D., Kamble, S., Huddar, N. and McNeil, G. (2018) 'Universals and cultural variations in 22 emotional expressions across five cultures', *Emotion*, *18*(1), 75–93.

Cottingham, J. (1996) 'General introduction: The *Meditations* and Cartesian philosophy', in René Descartes *Meditations on First Philosophy, with selections from the Objections and Replies*, edited and translated by J. Cottingham, pp. xviii–xxxviii. Cambridge: Cambridge University Press.

Cramer, J. S. (2007) *I to Myself: An Annotated Selection from the Journal of Henry D. Thoreau*. New Haven, CT: Yale University Press.

Crivelli, C. and Fridlund, A. (2018) 'Facial displays are tools for social influence', *Trends in Cognitive Sciences*, *22*(5), 388–99.

Crossley, N. (1995) 'Merleau-Ponty, the elusive body and carnal sociology', *Body & Society*, *1*(1), 43–63.

Crossley, N. (2005) 'Mapping reflexive body techniques: On body modification and maintenance', *Body & Society*, *11*(1), 1–35.

Crossley, N. (2006a) 'In the gym: Motives, meaning and moral careers', *Body & Society*, *12*(3), 23–50.

Crossley, N. (2006b) *Reflexive Sociology*. Maidenhead: Open University Press.

Czarniawska, B. (2007) *Shadowing and Other Techniques for Doing Fieldwork in Modern Societies*. Malmö: Liber.

Dale, K. (2001) *Anatomising Organisation Theory*. Basingstoke: Palgrave.

Damasio, A. (1994) *Descartes' Error: Emotion, Reason, and the Human Brain*. New York: Harper.

Damasio, A. (2003) *Looking for Spinoza: Joy, Sorrow and the Feeling Brain*. London: Heinemann.

Darwin, C. (1872) *The Expression of the Emotions in Man and Animals*. London: John Murray.

Day, K. (2010) 'Pro-anorexia and "binge-drinking": Conformity to damaging ideals or "new", resistant femininities?', *Feminism & Psychology*, *20*(2), 242–8.

de Beauvoir, S. (1995 [1976]) '*The Second Sex* 25 years later', interview with J. Gerassi, *Society*, *13* (January–February), 79–85.

de Beauvoir, S. (2011 [1949]) *The Second Sex*, translated by C. Borde and S. Malovany-Chevallier. New York: Vintage.

Deleuze, G. (1983 [1963]) *Kant's Critical Philosophy*, translated by H. Tomlinson and B. Habberjam. London: Athlone.

Deleuze, G. (1988 [1968]) *Bergsonism*, translated by H. Tomlinson and B. Habberjam. New York: Zone Books.

Deleuze, G. (1991 [1953]) *Empiricism and Subjectivity*, translated by C. Boundas. New York: Columbia University Press.

Deleuze, G. (1995 [1990]) *Negotiations: 1972–1990*. New York: Columbia University Press.

Deleuze, G. and Guattari, F. (1984 [1972]) *Anti-Oedipus: Capitalism and Schizophrenia, Vol. 1*, translated by H. R. Lane and R. Hurley. London: Athlone.

Deleuze, G. and Guattari, F. (1988 [1980]) *A Thousand Plateaus: Capitalism and Schizophrenia, Vol. 2*, translated by B. Massumi. London: Athlone.

DeLillo, D. (1993) 'Don DeLillo, The art of fiction No. 135', interviewed by A. Begley, *The Paris Review*, Issue 128 (Fall). www.theparisreview.org/interviews/1887/don-delillo-the-art-of-fiction-no-135-don-delillo

Delphy, C. and Leonard, D. (1992) *Familiar Exploitation: A New Analysis of Marriage in Contemporary Western Societies*. Cambridge: Polity Press.

de Rond, M. (2017) *Doctors at War: Life and Death in a Field Hospital*. Ithaca, NY: Cornell University Press.

Descartes, R. (1984) *The Philosophical Writings of Descartes, Vol. II*, translated by J. Cottingham, R. Stoothoff and D. Murdoch. Cambridge: Cambridge University Press.

Descartes, R. (1988) *The Philosophical Writings of Descartes*, 3 vols, translated by J. Cottingham, R. Stoothoff and D. Murdoch (*Vol. 3* incl. A. Kenny). Cambridge: Cambridge University Press.

Descartes, R. (1996 [1641]) *Meditations on First Philosophy, with selections from the Objections and Replies*, edited and translated by J. Cottingham. Cambridge: Cambridge University Press.

Descartes, R. (2007 [1637]) *Discourse on Method*, in the version by J. Bennett presented at www.earlymoderntexts.com

Descartes, R. (2012 [1644]) *Principles of Philosophy*, in the version by J. Bennett presented at www.earlymoderntexts.com

Dickens, C. (1838) *Nicholas Nickleby*. London: Chapman & Hall.

Dimberg, U., Thunberg, M. and Elmehed, K. (2000) 'Unconscious facial reactions to emotional facial expressions', *Psychological Science*, *11*(1), 86–9.

Douglas, M. (2001 [1966]) *Purity and Danger: An Analysis of Concepts of Pollution and Taboo*. London: Routledge & Kegan Paul.

Dreyfus, H. L. and Dreyfus, S. E. (1982) *Mind Over Machine*. New York: The Free Press.

Dreyfus, H. L. and Dreyfus, S. E. (2005) 'Peripheral vision: Expertise in real world contexts', *Organization Studies*, *26*(5): 779–92.

Duchenne de Bolougne, G. B. (1862) *Mécanisme de la Physionomie Humaine*. Paris: Jules Renouard.

Duncan, D. (2010) 'Embodying the gay self: Body image, reflexivity and embodied identity', *Health Sociology Review*, *19*(4), 437–50.

Durkheim, E. (1951 [1897]) *Suicide*, translated by J. Spaulding and G. Simpson. Glencoe, IL: Free Press.

Durkheim, E. (1973) *Emile Durkheim on Morality and Society*, edited by R. N. Bellah. Chicago: University of Chicago Press.

Durkheim, E. (1982 [1895]) *Rules of Sociological Method and Selected Texts on Sociology and its Method*, edited by S. Lukes, translated by W. D. Halls. New York: The Free Press.

Durkheim, E. (1982 [1909]) 'The contribution of sociology to psychology and philosophy', in *Rules of Sociological Method and Selected Texts on Sociology and its Method*, edited by S. Lukes, translated by W. D. Halls, pp. 236–40. New York: The Free Press.

Durkheim, E. (1984 [1893]) *The Division of Labor in Society*, translated by W. D. Halls. New York: The Free Press.

Efran, M. G. and Patterson, E. W. (1974) 'Voters vote beautiful: The effect of physical appearance on a national election', *Canadian Journal of Behavioral Science*, *6*(4), 352–6.

Ekman, P. (1992) 'An argument for basic emotions', *Cognition & Emotion*, *6*(3–4), 169–200.

Ekman, P. (1993) 'Facial expression and emotion', *American Psychologist*, *48*(4), 384–92.

Ekman, P. (2003) *Emotions Revealed: Understanding Faces and Feelings*. London: Weidenfeld & Nicolson.

Ekman, P. and Friesen, W. V. (1976) 'Measuring facial movement', *Environmental Psychology & Nonverbal Behavior*, *1*(1), 56–75.

Ekman, P. and Friesen, W. V. (1978) *Facial Action Coding System: A Technique for the Measurement of Facial Movement*. Palo Alto, CA: Consulting Psychologists Press.

Ekman, P., Friesen, W. V. and Ellsworth, P. (1972) *Emotion in the Human Face: Guidelines for Research and an Integration of Findings*. New York: Pergamon Press.

Ellis, C. (2016) 'Preface: Carrying the torch for autoethnography', in S. Holman Jones, T. E. Adams and C. Ellis (eds) *Handbook of Autoethnography*, pp. 9–12. Abingdon: Routledge.

Emmison, M., Smith, P. and Mayall, M. (2012) *Researching the Visual*, 2nd edition. London: Sage.

Engels, F. (1958 [1845]) *The Condition of the Working Class in England*, translated by W. O. Henderson and W. H. Chaloner. Oxford: Basil Blackwell.

Engelsrud, G. (2005) 'The lived body as experience and perspective: Methodological challenges', *Qualitative Research*, *5*(3), 267–84.

Fairclough, N. (2013) *Critical Discourse Analysis: The Critical Study of Language*. Abingdon: Routledge.

Fanon, F. (2008 [1952]) *Black Skin, White Masks*, translated by R. Philcox. New York: Grove Press.

Featherstone, M. (1982) 'The body in consumer culture', *Theory, Culture & Society*, *1*(1), 18–33.

Fine, G. (1993) *On Ideas: Aristotle's Criticism of Plato's Theory of Forms*. New York: Oxford University Press.

Flynn, M. (2011) 'Maurice Merleau-Ponty', in E. N. Zalta (ed.) *The Stanford Encyclopedia of Philosophy*. https://plato.stanford.edu/archives/spr2018/entries/merleau-ponty/

Foucault, M. (1973 [1963]) *The Birth of the Clinic: An Archaeology of Medical Perception*, translated by A. M. Sheridan. London: Tavistock.

Foucault, M. (1977 [1975]) *Discipline and Punish: The Birth of the Prison*, translated by A. Sheridan. London: Allen Lane.

Foucault, M. (1979 [1977]) *The History of Sexuality Volume 1: An Introduction*, translated by R. Hurley. Harmondsworth: Penguin.

Foucault, M. (1980) *Power/Knowledge: Selected Interviews and Other Writings 1972–1977 by Michel Foucault*, edited by C. Gordon, translated by C. Gordon, L. Marshall, J. Mepham and K. Soper. London: Harvester Wheatsheaf.

Foucault, M. (1990 [1966]) 'Maurice Blanchot: The thought from outside', translated by B. Massumi, in *Foucault–Blanchot*, pp. 7–58. New York: Zone.

Foucault, M. (2009 [1977–8]) *Security, Territory, Population*. New York: Picador.

Foucault, M. (2010 [1978–9]) *The Birth of Biopolitics*. New York: Picador.

Frank, A. (1997) *The Diary of a Young Girl: The Definitive Edition*, edited by O. H. Frank and M. Pressler, translated by S. Massotty. London: Penguin.

Fukuyama, F. (2018) *Identity: The Demand for Dignity and the Politics of Resentment*. New York: Farrar, Straus and Giroux.

Gailey, J. A. (2012) 'Fat shame to fat pride: Fat women's sexual and dating experiences', *Fat Studies*, *1*(1), 114–27.

Garfinkel, H. (1967) *Studies in Ethnomethodology*. New York: Prentice-Hall.

Gay, P. (2006 [1989]) *Freud: A Life of Our Time*. New York: Norton.

Geertz, C. (1973) *The Interpretation of Cultures*. New York: Basic Books.
Geertz, C. (1974) '"From the native's point of view": On the nature of anthropological understanding', *Bulletin of the American Academy of Arts & Sciences, 28*(1), 26–45.
Giddens, A. (1984) *Constitution of Society*. Cambridge: Cambridge University Press.
Gleason, J. (1989) *Special Education in Context: An Ethnographic Study of Persons with Developmental Disabilities*. Cambridge: Cambridge University Press.
Godfrey, R. and Brewis, J. (2018) '"Nowhere else sells bliss like this": Exploring the emotional labour of soldiers at war', *Gender, Work & Organization, 25*(6), 653–69.
Goehr, L. (2005) 'Understanding the engaged philosopher: On politics, philosophy and art', in T. Carman and M. B. N. Hansen (eds) *The Cambridge Companion to Merleau-Ponty*, pp. 318–51. Cambridge: Cambridge University Press.
Goffman, A. (2014) *One the Run: Fugitive Life in an American City*. Chicago: University of Chicago Press.
Goffman, E. (1959) *The Presentation of Self in Everyday Life*. New York: Anchor Books.
Goffman, E. (1961) *Encounters*. New York: Bobbs-Merrill.
Goldin-Meadow, S. and McNeill, D. (1999) 'The role of gesture and mimetic representation in making language the province of speech', in M. Corballis and S. Lea (eds) *The Descent of Mind*, pp. 155–72. Oxford: Oxford University Press.
Goldstein, B. M. (2007) 'All photos lie: Images as data', in G. C. Stanczak (ed.) *Visual Research Methods: Image, Society, and Representation*, pp. 61–81. Thousand Oaks, CA: Sage.
Greenland, S. and Rayman-Bacchus, L. (2014) 'Ethnography-photography: A visual approach to segmentation and living standard valuation in emerging markets', *Market & Social Research, 22*(1), 9–21.
Guest, R. (1955) 'Foremen at work – An interim report on method', *Human Organization, 14*(2), 21–4.
Guyer, P. and Wood, A. W. (1998) 'Introduction', in I. Kant, *Critique of Pure Reason*. Cambridge: Cambridge University Press.
Habermas, J. (1984 [1981]) *Theory of Communicative Action, Volume One: Reason and the Rationalization of Society*, translated by T. McCarthy. Boston: Beacon Press.
Halberstam, J. (1998) *Female Masculinity*. Durham, NC: Duke University Press.
Hamermesh, D. S. (2013) *Beauty Pays: Why Attractive People Are More Successful*. Princeton, NJ: Princeton University Press.
Hamington, M. (2008) 'Resources for feminist care ethics in Merleau-Ponty's phenomenology of the body', in G. Weiss (ed.) *Intertwinings: Interdisciplinary Encounters with Merleau-Ponty*, pp. 203–20. Albany, NY: SUNY Press.
Hampsten, E. (1989) 'Considering more than a single reader', in Personal Narratives Group (ed.) *Interpreting Women's Lives*, pp. 129–38. Bloomington, IN: Indiana University Press.
Hancock, P. and Tyler, M. (2000) 'The look of love: Gender and the organization of aesthetics', in J. Hassard, R. Holliday and H. Willmott (eds) *Body and Organization*, pp. 108–29. London: Sage.

Hanisch, C. (1970) 'The personal is political', in S. Firestone and A. Koedt (eds) *Notes From the Second Year: Women's Liberation*. New York (pamphlet).

Harari, Y. N. (2008) *The Ultimate Experience: Battlefield Revelations and the Making of Modern War Culture, 1450–2000*. Basingstoke: Macmillan.

Harari, Y. N. (2009) 'Scholars, eyewitnesses and flesh-witnesses of war: A tense relationship', *Partial Answers: A Journal of Literature and the History of Ideas*, *7*(2), 213–28.

Hardt, M. and Negri, A. (2000) *Empire*. Cambridge, MA: Harvard University Press.

Hardt, M. and Negri, A. (2004) *Multitude: War and Democracy in the Age of Empire*. New York: Penguin.

Harford, Tim (2018) 'Tim Harford's guide to statistics in a misleading age', *Financial Times*, 8 February. www.ft.com/content/ba4c734a-0b96-11e8-839d-41ca06376bf2?segmentId=080b04f5-af92-ae6f-0513-095d44fb3577

Haynes, K. (2012) 'Body beautiful? Gender, identity and the body in professional services firms', *Gender, Work & Organization*, *19*(5), 489–507.

Heath, C. (2004) 'Analysing face-to-face interaction: Video, the visual and material', in D. Silverman (ed.) *Qualitative Research: Theory, Method and Practice*, pp. 266–82. London: Sage.

Hedström, P. (2005) *Dissecting the Social: On the Principles of Analytical Sociology*. Cambridge: Cambridge University Press.

Heinämaa, S. (2003) *Toward a Phenomenology of Sexual Difference: Husserl, Merleau-Ponty, Beauvoir*. Lanham, MD: Rowman & Littlefield.

Hemingway, E. (1958) 'Ernest Hemingway, The art of fiction No. 21', interviewed by G. Plimpton, *The Paris Review*, Issue 18 (Spring). www.theparisreview.org/interviews/4825/ernest-hemingway-the-art-of-fiction-no-21-ernest-hemingway

Hengehold, L. (2007) *The Body Problematic: Political Imagination in Kant and Foucault*. University Park, PA: Penn State University Press.

Heyl, B. S. (2001) 'Ethnographic interviewing', in P. Atkinson, A. Coffey, S. Delamont, J. Lofland and L. Lofland (eds) *Handbook of Ethnography*, pp. 369–83. London: Sage.

Hindmarsh, J. and Pilnick, A. (2007) 'Knowing bodies at work: Embodiment and ephemeral teamwork in anaesthesia', *Organization Studies*, 28(9), 1395–416.

Hintikka, J. and Remes, U. (1974) *The Method of Analysis: Its Geometrical Origin and its General Significance*. Dordrecht: D. Reidel.

Hochschild, A. (1983) *The Managed Heart: Commercialization of Human Feeling*. Berkeley, CA: University of California Press.

Holman Jones, S., Adams, T. E. and Ellis, C. (2016) 'Introduction: Coming to know autoethnography as more than a method', in S. Holman Jones, T. E. Adams and C. Ellis (eds) *Handbook of Autoethnography*, pp. 17–47. Abingdon: Routledge.

hooks, b. (1986) 'Talking back', *Discourse*, *8*(Fall-Winter 86-87), 123–128.

Höpfl, H., Pullen, A. and Rhodes, C. (eds) (2015) 'Feminine writing of organizations', *Management Learning*, *22*(2), 87–196.

Horowitz, I. L. (1983) *C. Wright Mills: An American Utopian*. New York: Free Press.

Hughes, B. (2007) 'Being disabled: Towards a critical social ontology for disability studies', *Disability & Society, 22*(7), 673–84.

Hyde, M., Jappinen, P., Theorell, T. and Oxenstierna, G. (2006) 'Workplace conflict resolution and the health of employees in the Swedish and Finnish units of an industrial company', *Social Science & Medicine, 63*(8), 2218–27.

Imrie, R. (2004) 'Disability, embodiment and the meaning of the home', *Housing Studies, 19*(5), 745–763.

ipersonic (2017) 'The analytical thinker'. www.ipersonic.com/type/AT.html (accessed 14 December 2017).

Irigaray, L. (1985 [1977]) *This Sex Which Is Not One*, translated by C. Porter with C. Burke. Ithaca, NY: Cornell University Press.

Jacques, J. (2015) *Trans: A Memoir*. London: Verso.

Jorgensen, D. L. (1989) *Participant Observation: A Methodology for Human Studies*. Thousand Oaks, CA: Sage.

Kahnemann, D. (2011) *Thinking Fast and Slow*. New York: Farrar, Straus and Giroux.

Kant, I. (1998 [1770–87]) *Critique of Pure Reason*, translated and edited by P. Guyer and A. W. Wood. Cambridge: Cambridge University Press.

Kerfoot, D. and Knights, D. (1994) 'Power, identity and masculinity', in L. Radtke and H. Stam (eds) *Gender/Power: Social Relations in Theory and Practice*, pp. 78–89. London: Sage.

Kertzer, D. I. (2009) 'Social anthropology and social science history', *Social Science History, 33*(1), 1–16.

Kim, M., Roth, W. M. and Thom, J. (2011) 'Children's gestures and the embodied knowledge of geometry', *International Journal of Science & Mathematics Education, 9*(1), 207–38.

King, S. (2000) *On Writing: A Memoir of the Craft*. New York: Simon & Schuster.

King, S. (2006) 'Stephen King, The art of fiction No. 189', interviewed by C. Lehmann-Haupt and N. Rich, *The Paris Review*, Issue 178. www.theparisreview.org/interviews/5653/stephen-king-the-art-of-fiction-no-189-stephen-king

Knights D. and Clarke, C. (2014) 'A bittersweet symphony, this life: Fragile academic selves', *Organization Studies, 35*(3), 335–57.

Knights D. and Clarke, C. (2017) 'Pushing the boundaries of amnesia and myopia: A critical review of the literature on identity in management and organization studies', *International Journal of Management Reviews, 19*(3), 337–56.

Knights D. and Clarke, C. (2018) 'Living on the edge? Professions: their preoccupations and paranoias', *Culture & Organization, 24*(2), 134–53.

Knights, D. and Murray, F. (1994) *Managers Divided: Organisational Politics and Information Technology Management*. London: Wiley.

Knights, D. and Roberts, J. (1982) 'The power of organisation or the organisation of power? Management–staff relations in sales', *Organisation Studies, 3*(1), 47–63.

Kreuger, R. A. and Casey, M. A. (2015) *Focus Groups: A Practical Guide for Applied Research*, 5th edition. Thousand Oaks, CA: Sage.

Lakoff, G. and Johnson, M. (1999) *Philosophy in the Flesh: The Embodied Mind and its Challenge to Western Thought*. New York: Basic Books.

Landsberger, H. A. (1958) *Hawthorne Revisited: Management and the Worker: Its Critics, and Developments in Human Relations in Industry*. Ithaca, NY: Cornell University Press.

Laywine, A. (1993) 'Kant's early metaphysics and the origins of the critical philosophy', in The North American Kant Society, *Studies in Philosophy, Vol. 3*. Atascadero, CA: Ridgeview Publishing Company.

Le Guin, U. (2017) 'Ursula Le Guin on fiction vs. !'alternative facts': Letter to the editor', *The Oregonian*, 1 February. https://www.oregonlive.com/opinion/index.ssf/2017/02/ursula_leguin_on_fiction_vs_al.html

Leder, D. (1990) *The Absent Body*. Chicago: University of Chicago Press.

Lefort, C. (2005) 'Thinking politics', in T. Carman and M. B. N. Hansen (eds) *The Cambridge Companion to Merleau-Ponty*, pp. 352–79. Cambridge: Cambridge University Press.

Leibniz, G. W. (1989) *Philosophical Essays*, translated by R. Ariew and D. Garber. Indianapolis, IN: Hackett.

Levin, D. M. (1985) 'The body politic: Political economy and the human body', *Human Studies*, *8*(3), 235–78.

Lewthwaite, S. and Bartlett, R. (2017) 'Diary methods: Voyages into the interior', *Methods News: Newsletter from the ESRC National Centre for Research Methods*, *2*, 2.

Lilla, M. (2017) *The Once and Future Liberal: After Identity Politics*. New York: HarperCollins.

Linstead, S. and Pullen, A. (2006) 'Gender as multiplicity: Desire, displacement, difference and dispersion', *Human Relations*, *59*(9), 1287–1310.

Lipset, D. (1980) *Gregory Bateson: Legacy of a Scientist*. New York: Prentice-Hall.

Littlebear, N. (1983) 'Dreams of violence', in C. Moraga and G. Anzaldúa (eds) *This Bridge Called My Back: Writings by Radical Women of Color*, 2nd edition, pp. 16–17. New York: Kitchen Table/Women of Color Press.

Livingston, R. W. and Pearce, N. A. (2009) 'The teddy-bear effect: Does having a baby face benefit black Chief Executive Officers?', *Psychological Science*, *20*(10), 1229–36.

Lloyd, M. (1996) 'Feminism, aerobics and the politics of the body', *Body & Society*, *2*(2), 79–98.

Look, B. C. (2013) 'Leibniz's modal metaphysics', in E. N. Zalta (ed.) *The Stanford Encyclopedia of Philosophy*, spring 2013 edition. https://plato.stanford.edu/archives/spr2013/entries/leibniz-modal/

Lorenz, L. S. (2011) 'A way into empathy: A "case"of photo-elicitation in illness research', *Health*, *15*(3), 259–275.

Loveday, T. L. and Forster, E. S. (1913) *The Works of Aristotle*. Oxford: Clarendon Press.

Lu, J. L. (2009) 'Effect of work intensification and work extensification on women's health in the globalised labour market', *Journal of International Women's Studies, 10*(4), 111–26.

Lukes, S. (1982) 'Introduction', in E. Durkheim, *Rules of Sociological Method and Selected Texts on Sociology and its Method*, pp. 1–27. New York: The Free Press.

Lupton, D. (1995) *The Imperative of Health: Public Health and the Regulated Body*. London: Routledge.

MacDougal, D. (2005) *The Corporeal Image: Film, Ethnography and the Senses*. Princeton, NJ: Princeton University Press

Malinowski, B. (1961 [1922]) *Argonauts of the Western Pacific: An Account of Native Enterprise and Adventure in the Archipelagoes of Melanesian New Guinea*. New York: Dutton.

Malinowski, B. (1988 [1967]) *A Diary in the Strict Sense of the Term*, 2nd edition. London: Athlone.

Mangen, A., Walgermo, B. R. and Brønnick, K. (2013) 'Reading linear texts on paper versus computer screen: Effects on reading comprehension', *International Journal of Educational Research, 58*(1), 61–68.

Marx, K. (1973 [1857–8]) *Grundrisse: Foundations of the Critique of Political Economy*, translated by M. Nicolaus. Harmondsworth: Penguin.

Marx, K. (1976 [1867]) *Capital: A Critique of Political Economy, Vol. 1*, translated by B. Fowkes. London: Pelican.

Marx, K. (2009 [1959]) *Economic and Philosophic Manuscripts of 1844*, translated by M. Milligan, revised by D. J. Struik, corrected by M. Carmody. www.marxists.org/archive/marx/works/download/pdf/Economic-Philosophic-Manuscripts-1844.pdf

Mavin, S. and Grandy, G. (2016) 'Women elite leaders doing respectable business femininity: How privilege is conferred, contested and defended through the body', *Gender, Work & Organization, 23*(4), 379–96.

Mayo, E. (1945) *The Social Problems of an Industrial Civilization*. Boston: Graduate School of Business Administration, Harvard University.

McCracken, G. (1988) *The Long Interview*. Thousand Oaks, CA: Sage.

Merleau-Ponty, M. (1952) 'Les voix du silence', *Les Temps Modernes, 7* (July), 70–94.

Merleau-Ponty, M. (1962 [1945]) *The Phenomenology of Perception*, translated by C. Smith. London: Routledge & Kegan Paul.

Merleau-Ponty, M. (1964 [1960] *Signs*, translated by R. McCleary. Evanston, IL: Northwestern University Press.

Merleau-Ponty, M. (1968 [1964]) *The Visible and the Invisible*, translated by A. Lingis. Evanston, IL: Northwestern University Press.

Merleau-Ponty, M. (1969 [1947]) *Humanism and Terror: An Essay on the Communist Problem*, translated by J. O'Neill. Boston: Beacon Press.

Merleau-Ponty, M. (1973 [1955]) *Adventures of the Dialectic*, translated by J. Bien. Evanston, IL: Northwestern University Press.

Michaels, W. B. (2006) *The Trouble with Diversity: How We Learned to Love Identity and Ignore Inequality*. New York: Metropolitan.

Mills, C. W. (1959) *The Sociological Imagination*. New York: Oxford University Press.

Misra, J. and Murray-Close, M. (2014) 'The gender wage gap in the United States and cross nationally', *Sociology Compass*, *8*(11), 1281–1295.

Montag, C. and Reuter, M. (eds) (2015) *Internet Addiction: Neuroscientific Approaches and Therapeutical Interventions*. Cham: Springer:

Montag, C., Sindermann, C., Becker, B. and Panksepp, J. (2016) 'An affective neuroscience framework for the molecular study of internet addiction', *Frontiers in Psychology*, *7*, 1906–19.

Moraga, C. and Anzaldúa, G. (eds) (1983) *This Bridge Called My Back: Writings by Radical Women of Color*, 2nd edition. New York: Kitchen Table/Women of Color Press.

Morgan, K. P. (1991) 'Women and the knife: Cosmetic surgery and the colonization of women's bodies', *Hypatia*, *6*(3), 25–53.

Munro, A. (1994) 'Alice Munro, The art of fiction No. 137', interviewed by J. McCulloch and M. Simpson, *The Paris Review*, Issue 131 (Summer). www.theparisreview.org/interviews/1791/alice-munro-the-art-of-fiction-no-137-alice-munro

Munro, I. and Thanem, T. (2018a) 'Deleuze and the deterritorialization of strategy', *Critical Perspectives on Accounting*, *53*(1), 69–78.

Munro, I. and Thanem, T. (2018b) 'The ethics of affective leadership: Organizing good encounters without leaders', *Business Ethics Quarterly*, *28*(1), 51–69.

Myers, I. B. with Myers, P. B. (1995 [1980]) *Gifts Differing: Understanding Personality Type*. Mountain View, CA: Davies-Black Publishing.

Na, Y. and Choo, Y. H. (2011) 'Becoming a female-to-male transgender (FTM) in South Korea', in C. Bobel and S. Kwan (eds) *Embodying Resistance: Challenging the Norms, Breaking the Rules*, pp. 48–57. Nashville, TN: Vanderbilt University Press.

Neu, J. (1971) 'Plato's analogy of state and individual: The *Republic* and the organic theory of the state', *Philosophy*, *46*(177), 238–54.

New York Times (2017) 'Women's march highlights as huge crowds protest Trump: "We're not going away"', 21 January. www.nytimes.com/2017/01/21/us/womens-march.html

Newman, L. (2016) 'Descartes' epistemology', in E. N. Zalta (ed.) *The Stanford Encyclopedia of Philosophy*, winter 2016 edition. https://plato.stanford.edu/archives/win2016/entries/descartes-epistemology/

Newton, I. (1730) *Opticks: Or, A Treatise of the Reflections, Refractions, Inflections and Colours of Light*, 4th edition. London: Printed for William Innys at the West-End of St. Paul's.

Ngo, H. (2016) 'Racist habits: A phenomenological analysis of racism and the habitual body', *Philosophy & Social Criticism*, *42*(9), 847–72.

Nishio, M. (2013) 'Dickens's artistry of reporting verbs', online proceedings of the Annual Conference of the Poetics and Linguistics Association (PALA). www.pala.ac.uk/uploads/2/5/1/0/25105678/nishio.pdf (accessed 16 May 2018).

Olkowski, D. and Weiss, G. (eds) (2006) *Feminist Interpretations of Merleau-Ponty*. University Park, PA: Penn State University Press.
Orwell, G. (1933) *Down and Out in Paris and London*. London: Victor Gollancz.
Palmquist, S. R. (1993) *Kant's System of Perspectives: An Architectonic Interpretation of the Critical Philosophy*. Lanham, MD: University Press of America.
Palmquist, S. R. (2016) 'Kant's perspectival solution to the mind–body problem', *Culture & Dialogue, 4*(1), 194–213.
Parkins, W. (2000) 'Protesting like a girl: Embodiment, dissent and feminist agency', *Feminist Theory, 1*(1), 59–78.
Parsons, T. (1937) *The Structure of Social Action*. New York: The Free Press.
Parsons, T. (1951) *The Social System*. New York. The Free Press.
Paterson, K. and Hughes, B. (1999) 'Disability studies and phenomenology: The carnal politics of everyday life', *Disability & Society*, 14(5), 597–610.
Pennycook, G., Fugelsang, J. A. and Koehler, D. J. (2015) 'Everyday consequences of analytic thinking', *Current Directions in Psychological Science, 24*(6), 425–32.
Pepys, S. (2003) *The Diary of Samuel Pepys: A Selection*, selected and edited by R. Latham. London: Penguin.
Petersen, A. (1998) 'Sexing the body: Representations of sex differences in Gray's Anatomy, 1858 to the present', *Body & Society, 4*(1), 1–15.
Phillips, M., Pullen, A. and Rhodes, C. (2014) 'Writing organization as gendered practice: Interrupting the libidinal economy', *Organization Studies, 5*(3), 313–33.
Phillips, N. and Hardy, C. (2002) *Discourse Analysis: Investigating Processes of Social Construction*. London: Sage.
Pink, S. (2015) *Doing Sensory Ethnography*, 2nd edition. London: Sage.
Plato (1955) *Phaedo*, translated and edited by R. Hackforth. Cambridge: Cambridge University Press.
Plato (1987) *The Republic*, translated with an introduction by D. Lee. Harmondsworth: Penguin.
Plato (1997) *Complete Works*, edited by J. M. Cooper. Indianapolis, IN: Hackett.
Pope, H., Phillips, K. A. and Olivardia, R. (2000). *The Adonis Complex: The Secret Crisis of Male Body Obsession*. New York: Simon & Schuster.
Popper, K. (1962) *The Open Society and its Enemies*. London: Routledge & Kegan Paul.
Probyn, E. (2004) 'Everyday shame', *Cultural Studies, 18*(2–3), 328–49.
Protevi, J. (2001) *Political Physics*. London: Athlone.
Proust, M. (2016 [1913–27]) *Remembrance of Things Past*, translated by C. K. Scott Moncrieff (3 vols). Harmondsworth: Penguin.
Puggioni, R. (2014) 'Speaking through the body: Detention and bodily resistance in Italy', *Citizenship Studies, 18*(5), 562–77.
Pullen, A. (2018) 'Writing as labiaplasty', *Organization, 25*(1), 123–30.
Radley, A. (1995) 'The elusory body and social constructionist theory', *Body & Society, 1*(2), 3–23.

Ray, J. L. and Smith, A. D. (2012) 'Using photographs to research organizations: Evidence, considerations, and application in a field study,' *Organizational Research Methods, 15*(2), 288–315.

Reestorff, C. M. (2014) 'Mediatised affective activism: The activist imaginary and the topless body in the Femen movement', *Convergence: The International Journal of Research into New Media Technologies, 20*(4), 478–95.

Rhodes, C. (2018) 'Sense-ational organization theory! Practices of democratic scriptology', *Management Learning,* Online First. https://doi.org/10.1177/1350507618800716

Riach, K. and Warren, S. (2015) 'Smell organization: Bodies and corporeal porosity in office work', *Human Relations, 68*(5), 789–809.

Rich, A. (1976) *Of Woman Born: Motherhood as Experience and Institution.* London: Virago.

Rodis-Lewis, G. (1992) 'Descartes' life and the development of his philosophy', in J. Cottingham (ed.) *The Cambridge Companion on Descartes,* pp. 21–57. Cambridge: Cambridge University Press.

Rogers, J. K. and Henson, K. D. (1997) ' "Hey, why don't you wear a shorter skirt?" Structural vulnerability and the organization of sexual harassment in temporary clerical employment', *Gender & Society, 11*(2), 215–37.

Rose, G. (2016) *Visual Methodologies: An Introduction to Researching with Visual Materials.* London: Sage.

Rowles, G. D. (1978) *Prisoners of Space? Exploring the Geographical Experience of Older People.* Boulder, CO: Westview Press.

Russell, J. A. and Fernández-Dols, J. M. (1997) 'What does a facial expression mean?', in J. A. Russell and J. M. Fernández-Dols (eds) *The Psychology of Facial Expressions,* pp. 3–30. Cambridge: Cambridge University Press.

Scarry, E. (1985) *The Body in Pain.* Oxford: Oxford University Press.

Schilt, K. and Connell, C. (2007) 'Do workplace gender transitions make gender trouble?', *Gender, Work & Organization, 14*(6), 596–618.

Schwanen, T., Hardill, I. and Lucas, S. (2012) 'Spatialities of ageing: The co-construction and co-evolution of old age and space', *Geoforum, 43*(6), 1291–5.

Sclavi, M. (1989) *A una spanna da terra: Una giornata di scuola negli Stati Uniti e in Italia e i fondamenti di una metodologia umoristica.* Milan: Feltrinelli.

Seale, C. F. (1998) 'Qualitative interviewing', in C. F. Seale (ed.) *Researching Society and Culture.* London: Sage.

Shell, S. M. (1995) *The Embodiment of Reason: Kant on Spirit, Generation, and Community.* Chicago: Chicago University Press.

Shilling, C. (1993) *The Body and Social Theory.* London: Sage.

Shusterman, R. (2005) 'The silent, limping body of philosophy', in T. Carman and M. B. N. Hansen (eds) *The Cambridge Companion to Merleau-Ponty,* pp. 151–80. Cambridge: Cambridge University Press.

Silverman, D. (ed.) (2004) *Qualitative Research: Theory, Method and Practice*. London: Sage.

Silverman, D. (2006) *Interpreting Qualitative Data*, 3rd edition. London: Sage.

Silverman, E. K. (2005) 'Margaret Mead and Gregory Bateson in the Sepik, 1938: A timely polemic from a lost anthropological efflorescence', *Pacific Studies*, *28*(3/4), 128–41.

Slutskaya, N., Simpson, A. and Hughes, J. (2012) 'Lessons from photoelicitation: Encouraging working men to speak', *Qualitative Research in Organizations & Management: An International Journal*, *7*(1), 16–33.

Smith, S. (2008) 'The politics of identity', *International Socialist Review*, Issue 57 (January–February). www.isreview.org/issues/57/feat-identity.shtml

Spicker, S. (1970) 'Introduction', in S. Spicker (ed.) *The Philosophy of the Body: Rejections of Cartesian Dualism*, pp. 3–23. New York: Quadrangle/The New York Times Book Co.

Spinoza, B. de (1994 [1677]) *The Ethics and Other Works*, translated and edited by E. Curley. Princeton, NJ: Princeton University Press.

Spradley, J. P. (1980) *Participant Observation*. Orlando, FL: Harcourt College Publishers.

Stoller, P. (1997) *Sensuous Scholarship*. Philadelphia: University of Pennsylvania Press.

Straus E. W. (1966) 'The upright posture', in M. Natanson (ed.) *Essays in Phenomenology*, pp. 164–92. Dordrecht: Springer.

Sutton, B. (2007) 'Naked protest: Memories of bodies and resistance at the World Social Forum', *Journal of International Women's Studies*, *8*(3), 139–48.

Svare, H. (2006) *Body and Practice in Kant*. Dordrecht: Springer.

Sweetman, P. (1999) 'Anchoring the (postmodern) self? Body modification, fashion and identity', *Body & Society*, *5*(2–3), 51–76.

Taber, C. S. and Lodge, M. (2006) 'Motivated skepticism in the evaluation of political beliefs', *American Journal of Political Science*, *50*(3), 755–69.

Taylor, C. (2005) 'Merleau-Ponty and the epistemological picture', in T. Carman and M. B. N. Hansen (eds) *The Cambridge Companion to Merleau-Ponty*, pp. 26–49. Cambridge: Cambridge University Press.

Thanem, T. (2004) 'The body without organs: Nonorganizational desire in organizational life', *Culture & Organization*, *10*(3), 203–17.

Thanem, T. (2006) 'Living on the edge: Towards a monstrous organization theory', *Organization*, *13*(2), 163–93.

Thanem, T. (2010) 'Free at last? Assembling, producing and organizing sexual spaces in Swedish sex education', *Gender, Work & Organization*, *17*(1), 91–112.

Thanem, T. and Knights, D. (2012) 'Feeling and speaking through our gendered bodies: Embodied self-reflection and research practice in organisation studies', *International Journal of Work Organisation & Emotion*, *5*(1), 91–108.

Thanem, T. and Wallenberg, L. (2015) 'What can bodies do? Reading Spinoza for an affective ethics of organizational life', *Organization*, *22*(2), 235–50.

Thanem, T. and Wallenberg, L. (2016) 'Just doing gender? Transvestism and the power of underdoing gender in everyday life and work', *Organization*, *23*(2): 250–71.

Theorell, T. and Karasek, R. A. (1996) 'Current issues relating to psychosocial job strain and cardiovascular disease research', *Journal of Occupational Health Psychology*, *1*(1), 9–26.

Toombs, S. K. (2001) 'Reflections on bodily change: The lived experience of disability', in S. K. Toombs (ed.) *Handbook of Phenomenology and Medicine*, pp. 247–61. Dordrecht: Springer.

Tourish, D. (2015) 'It's time to retire the ABS guide to journal quality', *Times Higher Education*, 19 March, p. 32. www.timeshighereducation.com/comment/opinion/its-time-to-retire-the-abs-guide-to-journal-quality/2019112.article

Tressell, R. (1914) *The Ragged-Trousered Philanthropists*. London: Grant Richards.

Trethewey, A. (1999) 'Disciplined bodies: Women's embodied identities at work', *Organization Studies*, *20*(3), 423–50.

Turner, B. S. (1984) *The Body and Society*. London: Sage.

Turner, B. S. (1992) *Regulating Bodies: Essays in Medical Sociology*. London: Routledge.

Tutchell, E. and Edmonds, J. (2015) *Man-Made: Why So Few Women Are in Positions of Power*. Aldershot: Gower.

Tyler, M. and Abbott, P. (1998) 'Chocs away: Weight watching in the contemporary airline industry', *Sociology*, *32*(3), 433–50.

USA Today (2017) 'At 2.6 million strong, Women's marches crush expectations', 21 January. www.usatoday.com/story/news/politics/2017/01/21/womens-march-aims-start-movement-trump-inauguration/96864158/

Vachhani, S. (2012) 'The subordination of the feminine? Developing a critical feminist approach to the psychoanalysis of organizations', *Organization Studies*, *33*(9), 1237–55.

Van Maanen, J. (2011) 'Ethnography as work: Some rules of engagement', *Journal of Management Studies*, *48*(1), 218–34.

Vonnegut, K. (2012) *Letters*, edited by D. Wakefield. New York: Delacorte Press.

Wacquant, L. (1995) 'The pugilistic point of view: How boxers think and feel about their trade', *Theory & Society*, *24*(4), 489–535.

Wacquant, L. (2004) *Body and Soul: Notebook of an Apprentice Boxer*. New York: Oxford University Press.

Wacquant, L. (2015) 'For a sociology of flesh and blood', *Qualitative Sociology*, *38*(1), 1–11.

Wakefield, D. (2009) 'C. Wright Mills: Before his time', *The Nation*, 19 March. www.thenation.com/article/c-wright-mills-his-time/

Wall Street Journal (2018) 'What your CEO is reading: Developing talent; pumped to work; eating towards empathy', 16 March. https://blogs.wsj.com/cio/2018/03/16/

what-your-ceo-is-reading-developing-talent-pumped-to-work-eating-towards-empathy/

Warhurst, C. and Nickson, D. (2009) 'Who's got the look?' Emotional, aesthetic and sexualized labour in interactive services', *Gender, Work & Organization, 16*(3), 385–404.

Warren, S. (2002) 'Show me how it feels to work here: The role of photography in researching organizational aesthetics', *Ephemera: Theory & Politics in Organization, 2*(3), 224–45.

Watson, L. C. and Watson-Franke, M. B. (1985) *Interpreting Life Histories: An Anthropological Inquiry*. New Brunswick, NJ: Rutgers University Press.

West, C. and Zimmerman, D. H. (1987) 'Doing gender', *Gender & Society, 1*(2), 125–51.

Whitehead, A. N. (1967 [1925]) *Science and the Modern World*. New York: The Free Press.

Whyte, W. H. (1979) *The Social Life of Small Urban Spaces*. Documentary film, 58 mins, Direct Cinema Limited.

Whyte, W. H. (1980) *The Social Life of Small Urban Spaces*. Washington, DC: The Conservation Foundation.

Williams, S. J. and Bendelow, G. (1998) *The Lived Body: Sociological Themes, Embodied Issues*. London: Routledge.

Wolkowitz, C. (2006) *Bodies at Work*. London: Sage.

Wolkowitz, C. and Warhurst, C. (2010) 'Embodying labour', in P. Thompson and C. Smith (eds) *Working Life*, pp. 223–43. London: Palgrave.

Wong, E. M., Ormiston, M. E. and Haselhuhn, M. P. (2011) 'A face only an investor could love: CEOs' facial structure predicts their firms' financial performance', *Psychological Science, 22*(12), 1478–83.

Wood, A., Rychlowska, M., Korb, S. and Niedenthal, P. (2016) 'Fashioning the face: Sensorimotor simulation contributes to facial expression recognition', *Trends in Cognitive Sciences, 20*(3), 227–40.

Woodward, K. (2008) 'Hanging out and hanging about: Insider/outsider research in the sport of boxing', *Ethnography, 9*(4), 536–60.

Woolf, V. (1990) *A Moment's Liberty: The Shorter Diary*, abridged and edited by A. O. Bell. New York: Harcourt Brace.

Yakhlef, A. (2010) 'The corporeality of practice-based learning', *Organization Studies, 31*(4), 409–30.

Yanow, D. (2012) 'Organizational ethnography between toolbox and world-making', *Journal of Organizational Ethnography, 1*(1), 31–42.

Young, I. M. (1980) 'Throwing like a girl: A phenomenology of feminine body comportment, motility and spatiality', *Human Studies 3*(1), 137–156.

Young, I. M. (2005) *On Female Body Experience: 'Throwing Like a Girl' and Other Essays*. Oxford: Oxford University Press.

Young, L. and Barrett, H. (2001) 'Adapting visual methods: Action research with Kampala street children', *Area*, *33*(2), 141–52.

Zanker, C. and Gard, M. (2008) 'Fatness, fitness, and the moral universe of sport and physical activity', *Sociology of Sport Journal*, *25*(1), 48–65.

Žižek, S. (2017) *The Courage of Hopelessness: A Year of Acting Dangerously*. London: Melville House.

INDEX

NOTE: Page numbers followed by n. refer to notes.

Kolibri 360 GmbH
Ferdinand-Jühlke-Straße 7
99095 Erfurt, Deutschland
produktsicherheit@kolibri360.de